Ground Water in Freshwater-Saltwater Environments of the Atlantic Coast

by Paul M. Barlow

Circular 1262

U.S. Department of the Interior
U.S. Geological Survey

U.S. Department of the Interior
Gale A. Norton, Secretary

U.S. Geological Survey
Charles G. Groat, Director

U.S. Geological Survey, Reston, Virginia: 2003

For sale by U.S. Geological Survey, Information Services
Box 25286, Denver Federal Center
Denver, CO 80225

For more information about the USGS and its products:
Telephone: 1-888-ASK-USGS
World Wide Web: http://www.usgs.gov/

Any use of trade, product, or firm names in this publication is for descriptive purposes only and does not imply endorsement by the U.S. Government.

Although this report is in the public domain, it contains copyrighted materials that are noted in the text. Permission to reproduce those items must be secured from the individual copyright owners.

FOREWORD

Ground water is a vital resource for communities and ecosystems of the Atlantic coastal zone. Ground-water withdrawals for public supplies, agriculture, industry, and other uses in coastal counties from Maine to Florida were about 7.7 billion gallons per day in 1995, and in many coastal communities, ground water is the primary or sole source of drinking-water supply. Ground water sustains the flow of coastal streams and rivers and is a source of freshwater to coastal ponds, wetlands, and other coastal ecosystems. The U.S. Geological Survey (USGS) has a long history of conducting scientific studies of ground water in the Atlantic coastal zone. This Circular draws heavily on the body of scientific knowledge developed over several decades of USGS monitoring and studies.

Because of a growing awareness of the critical role of ground water in sustaining coastal populations, economies, and ecosystems, the time is right to review some of the important water-management issues and scientific principles related to ground water in the Atlantic coastal zone, and to identify some of the scientific and management challenges that lie ahead. This Circular describes the occurrence and flow of freshwater and saltwater in ground-water systems of the Atlantic coastal zone; reviews the causes, modes, and management of saltwater intrusion along the Atlantic coast; and illustrates some of the mechanisms of ground-water discharge and contaminant loading to coastal ecosystems. The Circular also illustrates traditional approaches for monitoring and managing saltwater intrusion, and highlights some of the innovative approaches being used to enhance the sustainability of coastal ground-water resources, such as desalination and aquifer storage and recovery systems. As coastal populations and ground-water use increase, new monitoring and research efforts will be needed to characterize the occurrence and movement of saline ground water in different types of coastal terrains and to better understand linkages between ground-water discharge and quality and the health of coastal ecosystems. The USGS looks forward to continued service to the citizens of the Atlantic coastal zone, providing science to help them manage their water resources.

Robert M. Hirsch
Associate Director for Water
U.S. Geological Survey

CONTENTS

Foreword iii

Introduction 1

Chapter 1. Occurrence and Flow of Freshwater and Saltwater in Coastal Aquifers 5

    Patterns of Freshwater-Saltwater Interaction in a Single-Layer Aquifer, Cape Cod, Massachusetts 16

    Effects of Sea-Level Fluctuations on the Freshwater-Saltwater Transition Zone, Northern Atlantic Coastal Plain Aquifer System 19

    Virginia's Inland Saltwater Wedge and the Chesapeake Bay Impact Crater 22

    Saltwater in the Floridan Aquifer System, South Carolina to Florida 26

Chapter 2. Causes, Modes, and Management of Saltwater Intrusion 31

    Development of a Desalination System in Response to Saltwater Intrusion, Cape May City, New Jersey 42

    Saltwater Intrusion from the Delaware River During Drought—Implications for the Effects of Sea-Level Rise on Coastal Aquifers 46

    Multifaceted Strategy for Managing Saltwater Intrusion in Coastal Georgia 49

    Vertical Migration of Saline Water Along Preferential Flow Conduits in the Floridan Aquifer System 56

    Vertical Migration of Saltwater Across Interconnected Aquifers in Water-Supply Wells, Florida 59

    Saltwater Intrusion in Southeastern Florida 61

Chapter 3. Detecting and Monitoring Saltwater Occurrence and Intrusion 70

    Saltwater-Monitoring Networks Along the Atlantic Coast 70

    Use of Strontium Isotopes to Identify Sources of Saline Ground Water, Southwestern Florida 76

    Application of Geophysical Methods in Coastal-Aquifer Studies 80

        Electromagnetic Methods to Delineate Freshwater-Saltwater Interfaces and Saltwater Intrusion 80

        Seismic-Reflection Surveys to Delineate Paleochannels, North Carolina to Georgia 85

Chapter 4. Ground Water and Coastal Ecosystems 87

    Ground-Water Discharge, Plant Distribution, and Nitrogen Uptake in a New England Salt Marsh 90

    Discharge of Nitrate-Contaminated Ground Water to a Coastal Estuary, Cape Cod, Massachusetts 96

    Submarine Ground-Water Discharge at Crescent Beach Spring, Florida 101

Challenges and Opportunities 104

Acknowledgments 106

References Cited 106

BOXES

Box A -- Chemical Characteristics and Sources of Saltwater **8**

Box B -- How Deep is it to Saltwater? **14**

Box C -- Saltwater Intrusion in a Fractured Crystalline-Rock Aquifer, Harpswell, Maine **34**

Box D -- Water-Level Response to Mandated Decreased Withdrawals in the New Jersey Coastal Plain **39**

Box E -- Numerical Modeling of Coastal Aquifers **54**

Box F -- Aquifer Storage and Recovery in South Florida **68**

Box G -- Robowell—An Automated Ground-Water Sampling Process to Remotely Monitor Saltwater Intrusion, Truro, Massachusetts **72**

Box H -- New Technologies to Track Ground-Water Flow and Nutrient Transport to Coastal Bays of Delaware and Maryland **88**

CONVERSION FACTORS, WATER-QUALITY UNITS, AND VERTICAL DATUMS

This report uses English and metric units. To determine equivalent metric values from English values, multiply the English values by the conversion factors listed below. To determine equivalent English values from metric values, divide the metric values by the conversion factors listed below.

Multiply	By	To Obtain
Length		
inch (in.)	25.4	millimeter (mm)
inch (in.)	2.54	centimeter (cm)
foot (ft)	0.3048	meter (m)
mile (mi)	1.609	kilometer (km)
Area		
acre	4,047	square meter (m^2)
square foot (ft^2)	0.09290	square meter (m^2)
square mile (mi^2)	2.590	square kilometer (km^2)
Volume		
gallon (gal)	3.785	liter (L)
cubic foot (ft^3)	0.02832	cubic meter (m^3)
Flow		
inch per year (in/yr)	25.4	millimeter per year (mm/yr)
foot per year (ft/yr)	0.3048	meter per year (m/yr)
cubic foot per second (ft^3/s)	0.02832	cubic meter per second (m^3/s)
gallon per minute (gal/min)	0.06308	liter per second (L/s)
million gallons per day (Mgal/d)	0.04381	cubic meter per second (m^3/s)
billion gallons per day (Bgal/d)	43.81	cubic meter per second (m^3/s)

Temperature is given in degree Celsius (°C), which can be converted to degree Fahrenheit (°F) by the following equation:
$$°F = 1.8\,(°C) + 32$$

Water-Quality Units

Abbreviations:
grams per cubic centimeter (g/cm^3)
milligrams per liter (mg/L)
parts per million (ppm)
parts per thousand (ppt)

Conversions: Most chemical concentrations in this report are given in milligrams per liter, which is a unit expressing the concentration of chemical constituents in solution as weight (milligrams) of solute per unit volume (liter) of water. A few of the chemical concentrations are given as parts per thousand or parts per million; these are units of weight of solute per weight of water. Parts per thousand (that is, grams of solute per kilogram of water) is a concentration that is often used in reporting the composition of seawater. Concentration expressed as parts per million (that is, milligrams of solute per kilogram of water) can be converted to milligrams per liter by multiplying the concentration by the density of water, in kilograms per liter. At low concentrations, such as that of freshwater, concentrations expressed as parts per million are nearly equal to those expressed as milligrams per liter.

Vertical Datums

Because this report is based on a large number of previously published scientific investigations, "sea level" is not referenced to a single vertical datum. "Mean sea level" also is not used with reference to a single datum; where used, the phrase means the average surface of the ocean as determined by calibration of measurements at tidal stations. The vertical datum used for each investigation described in this report is identified where it could be determined from the published sources of information.

Photograph courtesy of David Wilson, Maryland Coastal Bays Program

Ground Water in Freshwater-Saltwater Environments of the Atlantic Coast

by Paul M. Barlow

INTRODUCTION

The Nation's coastal regions are economically and environmentally vibrant areas that support major population centers and diverse ecosystems. Although they constitute less than 20 percent of the conterminous land area of the United States, coastal counties are home to more than half the Nation's population and many of its largest cities (National Oceanic and Atmospheric Administration, 1998). In 1995, more than 65 million people from Maine to Florida lived in coastal counties along the Atlantic Ocean and Gulf of Mexico, with large population centers within the Boston, Massachusetts, to Washington, D.C., corridor and in Florida (fig. 1A). Coastal areas include many of the fastest growing counties in the Nation, suggesting that coastal communities will continue to expand in the future.

Coastal populations and industries require a host of natural resources to sustain them, perhaps most important of which is a reliable source of freshwater. On a regional basis, the humid eastern seaboard has an abundant renewable freshwater supply, which has contributed to the growth and prosperity of the region. In 1995, total freshwater withdrawals for public supplies, agriculture, industry, and other offstream uses in coastal counties from Maine to Florida were about 30 billion gallons per day (Bgal/d), of which about 7.7 Bgal/d were supplied by ground water (fig. 1B) (U.S. Geological Survey, 2000). Although ground water supplied only about one-fourth of the total freshwater used, in many coastal communities ground water is the primary or sole source of drinking-water supply, and its use is increasing in many areas (fig. 2).

As ground-water use has increased in coastal areas, so has the recognition that ground-water supplies are vulnerable to overuse and contamination. Ground-water development depletes the amount of ground water in storage and causes reductions in ground-water discharge to streams, wetlands, and coastal estuaries and lowered water levels in ponds and lakes. Contamination of ground-water resources has resulted in degradation of some drinking-water supplies and coastal waters. Although overuse and contamination of ground water are not uncommon throughout the United States, the proximity of coastal aquifers to saltwater creates unique issues with respect to ground-water sustainability in coastal regions. These issues are primarily those of saltwater intrusion into freshwater aquifers and changes in the amount and quality of fresh ground-water discharge to coastal saltwater ecosystems.

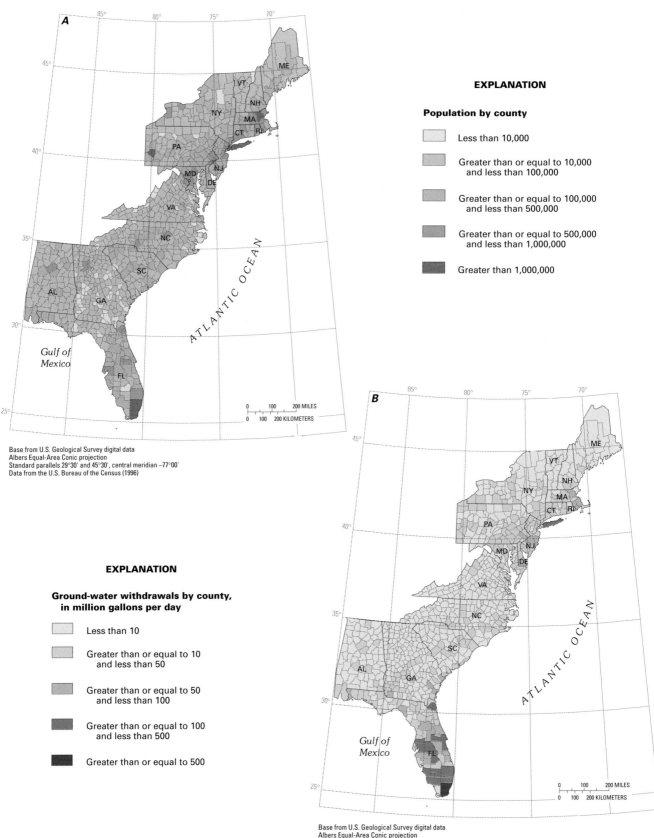

Figure 1. Population (A) and ground-water withdrawals (B) by county, 1995, in States along the Atlantic coast of the United States. More than two-thirds of the population of these States live in coastal counties, defined as having at least 15 percent of their land area in a coastal watershed (National Oceanic and Atmospheric Administration, 1998).

Saltwater intrusion is the movement of saline water into freshwater aquifers and most often is caused by ground-water pumping from coastal wells. Because saltwater has high concentrations of total dissolved solids and certain inorganic constituents, it is unfit for human consumption and many other anthropogenic uses. Saltwater intrusion reduces fresh ground-water storage and, in extreme cases, leads to the abandonment of supply wells when concentrations of dissolved ions exceed drinking-water standards. The problem of saltwater intrusion was recognized as early as 1854 on Long Island, New York (Back and Freeze, 1983), thus predating many other types of drinking-water contamination issues in the news.

Coastal ecosystems are sensitive to the salinity and nutrient concentrations of coastal waters. In recent years, scientists, coastal managers, and public decision-makers have recognized that many of the environmental issues related to coastal ecosystems—red tides, fish kills, loss of seagrass habitats, and destruction of coral reefs—can be attributed to the introduction of excess nutrients (nitrogen and phosphorus) from freshwater discharges (National Research Council, 2000). Ground water can be a significant source of freshwater to some coastal waters, and its role in delivering excess nutrients to coastal ecosystems is of increasing concern because of widespread nutrient contamination of shallow ground water (U.S. Geological Survey, 1999).

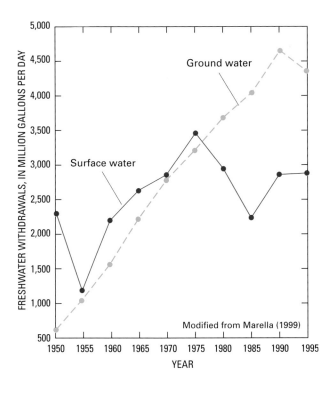

Figure 2. Ground-water withdrawals in Florida have increased sharply since 1950 and in 1995 accounted for 60 percent of total freshwater withdrawals in the State. Ninety-three percent of the 14.2 million people in Florida relied on ground water for their drinking-water needs in 1995 (Marella, 1999).

This report summarizes the current (2003) understanding of ground water in freshwater-saltwater environments of the Atlantic coast from Maine to Florida (including areas of Florida that are along the eastern Gulf of Mexico). These environments include coastal aquifers, salt marshes, and coastal waters (fig. 3). The report describes the natural occurrence and flow of freshwater and saltwater in coastal aquifers and the causes and modes of saltwater intrusion along the Atlantic coast. The role of ground water in coastal ecosystems—an area of increasing scientific and resource-management interest—also is reviewed. The report makes extensive use of case studies to illustrate the variety of freshwater-saltwater interactions that occur along the Atlantic coast—interactions that take place on time scales that range from daily tidal cycles to millennia and on spatial scales that range from beach faces a few feet wide to regional aquifer systems that extend over thousands of square miles. The case studies also demonstrate the many approaches and tools that are used to detect and monitor saltwater in coastal aquifers and to manage and prevent saltwater intrusion.

Although this report focuses on conditions along the Atlantic coastal zone, saltwater intrusion has been a problem in other coastal areas of the Nation and throughout the world (Bear and others, 1999). Moreover, it is now recognized that ground water can be an important global pathway for transport of materials from the land to coastal ecosystems and the oceans (for example, see Johannes, 1980; Simmons, 1992; Church, 1996; Moore, 1996, 1999; Basu and others, 2001; and Burnett and others, 2002).

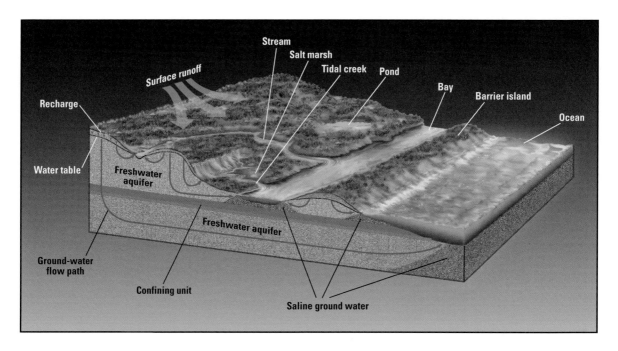

Figure 3. Ground-water flow paths in an idealized coastal watershed along the Atlantic coast. Fresh ground water is bounded by saline ground water beneath the bay and ocean. Fresh ground water discharges to coastal streams, ponds, salt marshes, and tidal creeks and directly to the bay and ocean.

CHAPTER 1. OCCURRENCE AND FLOW OF FRESHWATER AND SALTWATER IN COASTAL AQUIFERS

Ground water occurs in pores, fractures, solution cavities, and other openings in geologic formations that underlie the Atlantic coastal zone (fig. 4). The nature of the water-bearing openings within a specific geologic formation depends to a large extent on the mineral composition and structure of the formation and the geologic processes that initially formed and then further modified it. Geologic formations that occur along the Atlantic coast are of three major groups—unconsolidated (soil-like) deposits, semiconsolidated deposits, and consolidated rocks (Heath, 1998; Miller, 1999). Unconsolidated deposits consist of rock fragments and mineral grains that range in size from fractions of a millimeter (clay size) to several meters (boulders). In contrast, consolidated rocks consist of mineral particles that have been welded by heat and pressure or cemented by chemical reactions and precipitation into a solid mass. Consolidated rocks commonly are referred to as bedrock. Semiconsolidated deposits fall between the extremes of unconsolidated deposits and consolidated rocks (Heath, 1998). Although there are numerous types of geologic formations along the Atlantic coast, the most important water-bearing formations are unconsolidated sands and gravels; semiconsolidated sands; and, among the consolidated rocks, carbonates (primarily limestones), sandstones, and fractured crystalline rocks such as granites.

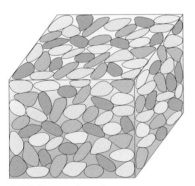
A. Well-sorted sand

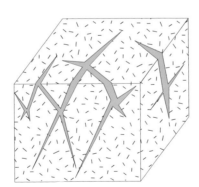
B. Fractures in granite

C. Caverns in limestone

Modified from Heath (1998)

Figure 4. Ground water occurs in the pore spaces of geologic formations. There are two types of porosity: primary porosity, which refers to openings that formed at the same time as the rock, such as the pores in well-sorted unconsolidated sand; and secondary porosity, which refers to openings that formed after the rock was formed, such as fractures in granites and solution cavities in limestone.

Saturated geologic materials that yield usable quantities of water to a well or spring are called aquifers. An aquifer can consist of a single geologic formation, a group of formations, or part of a formation (Lohman and others, 1972). The capacity of a geologic material to transit water is characterized by the material's hydraulic conductivity, which is commonly referred to as the permeability of the material. Confining units (or confining layers) are geologic units that are less permeable than aquifers. Because of their lower permeability, confining units restrict the movement of ground water into or out of adjoining aquifers. Aquifers and confining units are mapped on the basis of the degree of contrast in hydraulic conductivity among geologic units (Sun and Johnston, 1994). Generally, there is a close correlation between the type of geologic formation and its water-yielding properties. For example, unconsolidated sands and gravels, sandstones, and limestones commonly are major sources of ground-water supplies (aquifers), whereas beds of silt and clay function primarily as confining units (Heath, 1984).

Ground water in the Atlantic coastal zone occurs in confined and unconfined aquifers. Where water completely fills the pore spaces of an aquifer that is overlain by a confining unit, the aquifer is referred to as confined (or artesian). In contrast, where water only partially fills the pore spaces of an aquifer, the upper surface of the saturated zone (which is called the water table) is free to rise and decline, and the aquifer is referred to as unconfined (or as a water-table aquifer). Aquifers within the Atlantic coastal zone vary in size from local-scale aquifers that are a few square miles or less in areal extent to multilayer, regional-scale aquifers that are tens of thousands of square miles in areal extent. In many areas, unconfined aquifers that are close to land surface are underlain by one or more confined aquifers that may be partially or completely isolated from the land surface by confining units (fig. 5). Although these multilayer, regional aquifer systems may be discontinuous locally, they act hydrologically as a single system on a regional scale (Sun and Johnston, 1994).

The general pattern of fresh ground-water flow in coastal aquifers is from inland recharge areas where ground-water levels (hydraulic heads) typically are highest to coastal discharge areas where ground-water levels are lowest. This pattern of flow is illustrated by the ground-water flow paths in figures 3 and 5. Hydraulic head (often simply referred to as "head") is a measure of the total energy available to move ground water through an aquifer, and ground water flows from locations of higher head (that is, higher energy) to locations of lower head (lower energy). The distribution of hydraulic head within an aquifer is determined by measuring ground-water-level elevations in observation wells that are open to a small interval of the aquifer. The ground-water level (elevation) at each well most often is reported as feet (or meters) above or below sea level. Note that the upward direction of some of the ground-water flow paths near and beneath the ocean in the aquifer system shown in figure 5 indicates that ground-water heads in the deeper part of the flow system are above sea level near the coast.

Fresh ground water comes in contact with saline ground water at the seaward margins of coastal aquifers. The seaward limit of freshwater in a particular aquifer is controlled by the amount of freshwater flowing through the aquifer, the thickness and hydraulic properties of the aquifer and adjacent confining units, and the relative densities of saltwater and freshwater, among other variables. Because of its lower density, freshwater tends to remain above the saline (saltwater) zones of the aquifer, although in multilayered aquifer systems, seaward-flowing freshwater can discharge upward through confining units into overlying saltwater (fig. 5). As used in this report, saltwater is defined as water having a total dissolved-solids concentration greater than 1,000 milligrams per liter (mg/L) (Box A). Seawater has a total dissolved-solids concentration of about 35,000 mg/L, of which dissolved chloride is the largest component (about 19,000 mg/L). Concentrations of chloride in fresh ground water along the Atlantic coast are typically less than about 20 mg/L, so there is a large contrast in chloride concentrations between freshwater and saltwater.

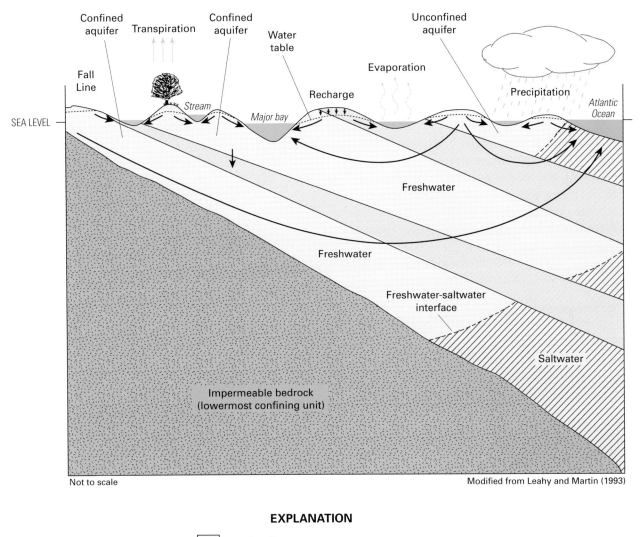

Figure 5. Generalized ground-water flow patterns in a multilayer, regional aquifer system. The water table forms the upper boundary to the uppermost aquifer in the sequence, which is called an unconfined (or water-table) aquifer. Although the two lower aquifers also are bounded by water tables near the land surface, the aquifers are considered to be confined because each of them is overlain and underlain by confining units. Recharge occurs to the upper, unconfined parts of the aquifer system. Ground water discharges to streams, rivers, and creeks, through the floors of the ocean and bays, and to marshes fringing coastal waters.

A

Chemical Characteristics and Sources of Saltwater

All water contains dissolved chemical materials called "salts." When the concentration of these dissolved materials becomes great, the water is referred to as "saltwater" or as "salty", "brackish", or "saline." These terms commonly are used interchangeably, which can be confusing because they can mean different things to different people. As a result, different classification methods have been developed to differentiate between freshwater and saltwater and to define the degree of salinity (that is, the total dissolved-solids concentration) of the water.

The classification method used in this report defines freshwater as having a total dissolved-solids concentration of less than 1,000 mg/L; waters with a total dissolved-solids concentration greater than 1,000 mg/L are considered to be saltwater (or saline). This somewhat arbitrary upper limit of freshwater is based on the suitability of the water for human consumption. Although waters with total dissolved-solids concentrations of greater than 1,000 mg/L have been used for domestic supply in some areas of the United States where water of lower dissolved-solids concentration is not available (Krieger and others, 1957; Feth and others, 1965), water containing more than 2,000 to 3,000 mg/L total dissolved solids is generally too salty to drink (Freeze and Cherry, 1979). Brackish waters can be defined as those having a total dissolved-solids concentration of 1,000 to 35,000 mg/L. The upper concentration limit for brackish water is set at the approximate concentration of seawater (35,000 mg/L). The average concentrations of the major dissolved constituents of seawater (those with concentrations exceeding 1.0 mg/L) are given in table A–1; chloride, sodium, sulfate, and magnesium have the largest concentrations. Water with a dissolved-solids concentration exceeding that of seawater is called a brine. Although there are different types of brines in terms of chemical composition, the largest number represent concentrated seawater containing mostly sodium chloride (Krieger and others, 1957).

The U.S. Environmental Protection Agency (USEPA) has established secondary maximum contaminant levels (SMCLs) for total dissolved solids, chloride, and sulfate in drinking water. Unlike maximum contaminant levels (MCLs) that have been established to protect the public against drinking-water contaminants that present a risk to human health, the secondary contaminant levels have been established as guidelines to assist operators of public-water systems in managing the aesthetic qualities of the water such as taste, color, and odor. The SMCL set by the USEPA for total dissolved solids is 500 mg/L,

Table A–1. Average concentrations of major dissolved constituents of seawater (from Hem, 1989, p. 7)

Constituent	Concentration (milligrams per liter)
Chloride	19,000
Sodium	10,500
Sulfate	2,700
Magnesium	1,350
Calcium	410
Potassium	390
Bicarbonate	142
Bromide	67
Strontium	8
Silica	6.4
Boron	4.5
Fluoride	1.3

whereas those for chloride and sulfate are each 250 mg/L (U.S. Environmental Protection Agency, 1992, 2002a). Concentrations of chloride in fresh ground water along the Atlantic coast are typically less than about 20 mg/L. Because of the high concentration of chloride in seawater (19,000 mg/L), less than a 2-percent contribution of seawater mixed with fresh ground water would render the water unsuitable for public supply using the USEPA guideline. The USEPA also has established a drinking-water advisory for sodium. The advisory recommends reducing sodium concentrations in drinking water to concentrations of 30 to 60 mg/L to avoid adverse effects on taste; a health-based guidance level of 20 mg/L in drinking water also is recommended by the USEPA for individuals on a very low sodium diet (less than 500 milligrams per day) (U.S. Environmental Protection Agency, 2002b). In some instances, States along the Atlantic coast have established standards for dissolved constituents in drinking water that differ from the levels established by the USEPA.

Two additional characteristics of water that are important in ground-water systems are density and viscosity, both of which are dependent on the type and amount of solutes dissolved in the water. As noted by Reilly (1993), density is important because it is part of the driving force that defines the direction and rate of fluid movement through a ground-water system; moreover, the density and viscosity of the water affect the hydraulic transmitting properties (permeability and hydraulic conductivity) of the ground-water system, which influence the rate of fluid movement.

Although seawater in the ocean and estuaries that laterally bound the eastern seaboard is by far the primary source of saline water to coastal ground-water systems, a number of other sources can affect coastal ground-water quality (Krieger and others, 1957; Feth and others, 1965; Task Committee on Saltwater Intrusion, 1969; Custodio, 1997; Richter and others, 1993; Jones and others, 1999). These sources include:

- Precipitation: Oceans are the largest single source of salts in the atmosphere, and sodium and chloride are the most abundant ions in air masses over the sea (Feth, 1981). Chloride and sodium concentrations, therefore, are high in air masses near sea coasts but decrease rapidly with increasing distance inland. These airborne salts are delivered to coastal watersheds by precipitation. Chloride concentrations in precipitation, however, are relatively small compared to seawater. Average chloride concentrations in precipitation measured at about two-dozen atmospheric deposition monitoring stations along the Atlantic coast in 2000, for example, ranged from less than 0.2 to 2.8 mg/L (National Atmospheric Deposition Program, 2001). Concentrations of sodium and chloride can be increased in soils, shallow surface waters (such as tidal lagoons), and ground water by evaporation and evapotranspiration.

- Sea-spray accumulation, tides, and storm surges, which can be local sources of increased ground-water salinity in low-lying coastal areas.

- Entrapped fossil seawater in unflushed parts of an aquifer: Such water either was trapped in sedimentary formations when they were deposited (connate water) or flowed into the formations during periods of relatively high sea levels when seawater flooded low-lying coastal areas.

- Dissolution of evaporitic deposits such as halite (rock salt), anhydrite, and gypsum (Manheim and Horn, 1968; Meisler, 1989).

- Pollution from various anthropogenic sources including sewage and some industrial effluents, oil- and gas-field brines brought to the land surface during exploration and production, road deicing salts, and return flows of irrigation water.

The freshwater and saltwater zones within coastal aquifers are separated by a transition zone (sometimes referred to as the zone of dispersion) within which there is mixing between freshwater and saltwater (figs. 6 and 7). The transition zone is characterized most commonly by measurements of either the total dissolved-solids concentration or of the chloride concentration of ground water sampled at observation wells. Although there are no standard practices for defining the transition zone, concentrations of total dissolved solids ranging from about 1,000 to 35,000 mg/L and of chloride ranging from about 250 to 19,000 mg/L are common indicators of the zone. In this report, the term "transition zone" implies a change in the quality of ground water from freshwater to saltwater, as measured by an increase in dissolved constituents such as total dissolved solids and chloride.

Within the transition zone, freshwater flowing to the ocean mixes with saltwater by the processes of dispersion and molecular diffusion. Mixing by dispersion is caused by spatial variations (heterogeneities) in the geologic structure and the hydraulic properties of an aquifer and by dynamic forces that operate over a range of time scales, including daily fluctuations in tide stages (fig. 8), seasonal and annual variations in ground-water recharge rates, and long-term changes in sea-level position (figs. 9 and 10). These dynamic forces cause the freshwater and saltwater zones to move seaward at times and landward at times. Because of the mixing of freshwater and saltwater within the transition zone, a circulation of saltwater is established in which some of the saltwater is entrained within the overlying freshwater and returned to the sea, which in turn causes additional saltwater to move landward toward the transition zone (fig. 6).

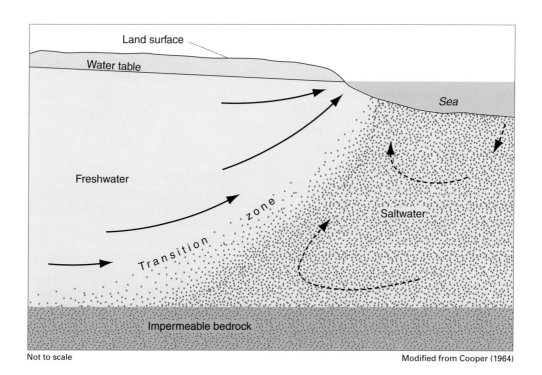

Figure 6. Ground-water flow patterns and the freshwater-saltwater transition zone in an idealized coastal aquifer. A circulation of saltwater from the sea to the transition zone and then back to the sea is induced by mixing of freshwater and saltwater in the transition zone.

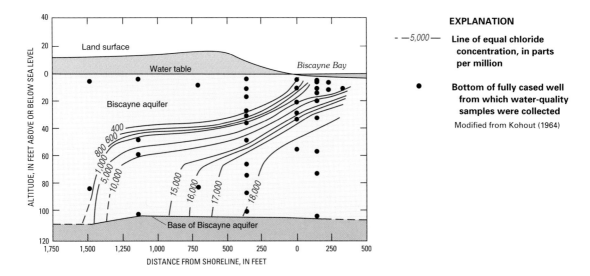

Figure 7. The transition zone in the Biscayne aquifer near Miami, Florida. The transition zone was determined by measurement of the chloride concentration of water samples extracted from the bottom of monitoring wells (shown as black dots). Chloride concentrations in the transition zone increased in the seaward direction from freshwater containing 16 parts per million to seawater containing 19,000 parts per million (Kohout, 1964).

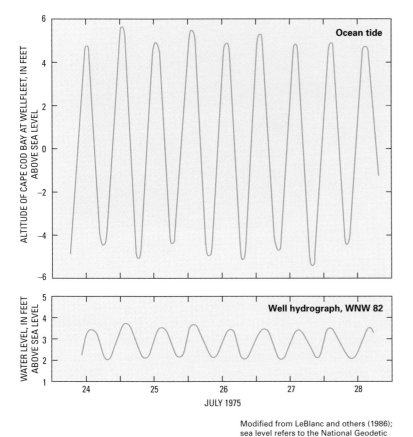

Figure 8. Mixing of freshwater and saltwater in the transition zone is caused in part by the forces of rising and falling ocean tides that continually push and pull the freshwater-saltwater interface first landward and then seaward during each tidal cycle. Tidal fluctuations also cause cyclic fluctuations of ground-water levels near the coast, as seen in the water-level hydrograph of well WNW 82 located 500 feet from the shoreline in Wellfleet, Massachusetts. The well is open to the aquifer from 37 to 40 feet below sea level. The tidal fluctuations are transmitted hydraulically into the freshwater aquifer and cause water-level fluctuations in the well that are of similar frequency but smaller amplitude. The effect of tidal fluctuations is most pronounced at the shoreline and decreases rapidly with increased distance inland from the coast. (Location of monitoring site shown in figure 11.)

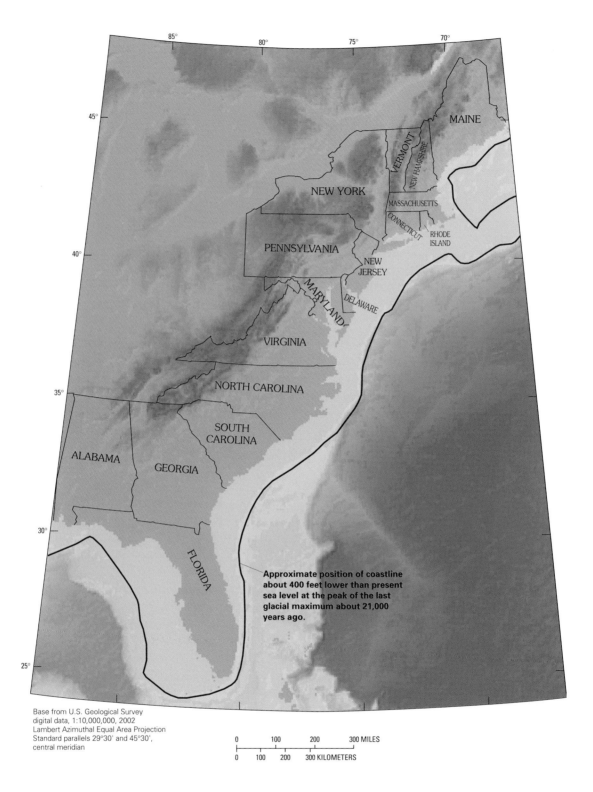

Figure 9. Large-scale movement and mixing of the freshwater-saltwater transition zone is caused in part by long-term fluctuations in global sea-level position. Sea-level transgressions onto the continent and subsequent regressions to positions seaward of the modern coastline have occurred several times in the geologic past. For example, the global sea level was a maximum of about 400 feet (125 meters) below the present level at the peak of the last glacial maximum about 21,000 years ago and has since risen to its present level (Fairbanks, 1989; Douglas and others, 2001). As the sea level declines, land areas are exposed to freshwater recharge, pushing the freshwater-saltwater interface seaward; as the sea level rises, the interface is pushed landward, and saltwater displaces freshwater zones of the aquifer.

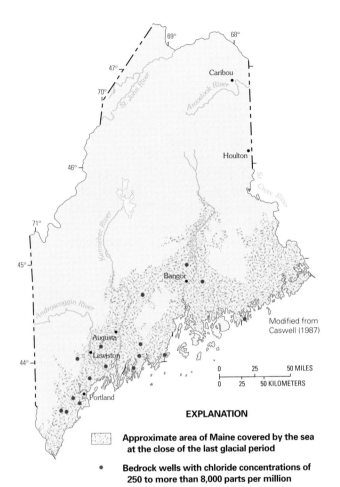

Figure 10. Along parts of the Atlantic coast from Maine to Massachusetts, broad lowland areas were flooded by the sea during the close of the last period of glaciation, approximately 12,000 to 13,000 years ago. This flooding occurred in areas that were temporarily depressed below sea level by the immense weight of the glacial ice (Olcott, 1995). As the glaciers retreated and before the land surface rebounded above the present sea level, there was a period during which seawater submerged these lowland areas, and saltwater intruded the unconsolidated sediments and fractured bedrock that formed the inland seafloor. After the land surface rebounded and the ocean receded, freshwater replaced this saline ground water in most areas. However, in some regions of relatively slow ground-water circulation, the saltwater has been trapped in the sediments and bedrock for about 12,000 years (Tepper, 1980; Snow, 1990). Several wells in Maine that are distant from the coast have yielded water with high chloride concentrations that have been attributed to this trapped seawater.

The horizontal (or lateral) width of the transition zone can be narrow, such as the approximately 1,500-foot (ft) width shown for the Biscayne aquifer in figure 7, or very wide, such as in parts of the New Jersey Coastal Plain where the transition zone spans several miles. The vertical thickness of the transition zone also varies among aquifers, but generally is much smaller than the horizontal width and is limited by the total thickness of the aquifer. For example, the thickness of the transition zone shown for the Biscayne aquifer ranges from about 50 to 70 ft (fig. 7). For the convenience of illustrating freshwater-saltwater interactions as simply as possible and facilitating simplified scientific analysis of these interactions when possible, the freshwater and saltwater zones often are assumed to be separated by a sharp boundary that is referred to as the freshwater-saltwater interface, such as those shown in figure 5. Although the depth to this interface is quite variable along the Atlantic coast, it can be estimated approximately under some circumstances by using a technique known as the Ghyben-Herzberg relation (Box B).

The variety of geologic settings, aquifer types, and hydrologic conditions along the Atlantic coast has resulted in many patterns of freshwater-saltwater flow and mixing in coastal aquifers. The four case studies that follow illustrate some of the important freshwater-saltwater environments that exist in aquifers along the Atlantic coast and highlight the many variables that control the natural occurrence and flow of freshwater and saltwater in coastal aquifers. The case studies progress from the glacial aquifer of Cape Cod, Massachusetts, which is representative of shallow, single-layer aquifers, to two of the most productive regional aquifer systems in the United States—the Northern Atlantic Coastal Plain aquifer system that extends from Long Island, New York, through North Carolina, and the Floridan aquifer system that extends from South Carolina to Alabama. These are thick, multilayered aquifer systems that underlie thousands of square miles.

How Deep is it to Saltwater?

Many who live along the coast often wonder how deep it is to salty ground water, particularly those searching for freshwater supplies. Although it is difficult to estimate the thickness of the freshwater zone in the absence of water-quality data, there is a relatively simple equation that has been used successfully in many areas to estimate the depth to saltwater and the thickness of freshwater in a water-table (unconfined) aquifer. The equation relates the elevation of the water table to the elevation of the boundary of the interface between the freshwater and underlying saltwater zones of an aquifer (fig. B–1), and is based on the balance of the height of two columns of fluids of differing density (Reilly and Goodman, 1985). In the equation, the thickness of the freshwater zone above sea level is represented as h and that below sea level is represented as z, as shown in figure B–1. The two thicknesses are related by

$$z = \frac{\rho_f}{\rho_s - \rho_f} h \qquad (1)$$

where

ρ_f is the density of freshwater, and

ρ_s is the density of saltwater.

This equation often is called the Ghyben-Herzberg relation after two European scientists who derived it independently in the late 1800s.

Freshwater has a density of about 1.000 grams per cubic centimeter (g/cm^3) at 20 °C, whereas that of seawater is about 1.025 g/cm^3. Although the difference between the density of freshwater and seawater is small, equation 1 indicates that this density contrast results in 40 ft of freshwater below sea level for every 1 ft of freshwater above sea level, that is:

$$z = 40h \qquad (2)$$

The total thickness of the freshwater zone is the sum of the freshwater zones above and below sea level: $h + z$. An example of the application of equations 1 and 2 is shown in figure B–2.

The Ghyben-Herzberg method of calculating the thickness of the freshwater zone in a coastal water-table aquifer is based on a number of simplifying assumptions. Most importantly, it is assumed that hydrostatic conditions exist within the aquifer, which implies that there are no vertical gradients in ground-water levels (heads) within the aquifer. This assumption incorrectly causes the thickness of the freshwater zone to be represented as zero at the shore where the elevation of the water table is zero. In reality, the freshwater zone must have some thickness for freshwater to discharge to the ocean. It also is assumed that there is an abrupt (or sharp) boundary between the freshwater and saltwater zones at the interface between the two types of water. However, as shown earlier, freshwater and saltwater mix at the boundary between the two waters, which results in a zone of dispersion at the interface.

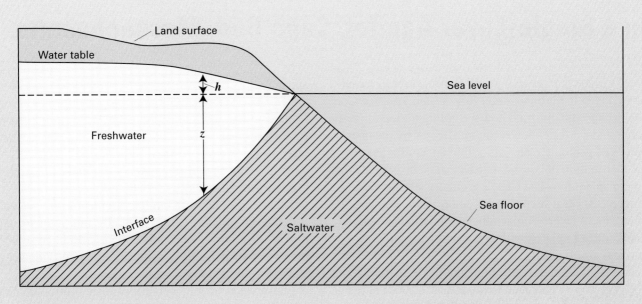

Figure B–1. *Simplified freshwater-saltwater interface in a coastal water-table aquifer.*

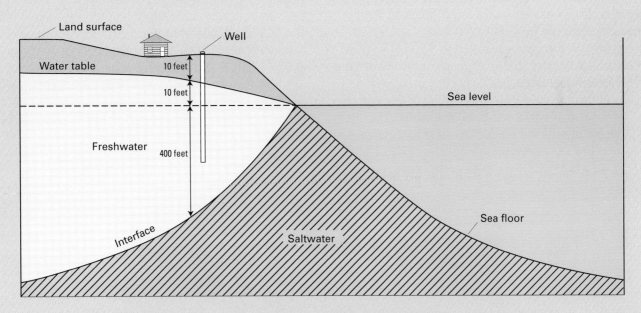

Figure B–2. *Example application of equations 1 and 2 for estimating the thickness of freshwater in a coastal water-table aquifer: A freshwater-supply well near the coast has deteriorated with time and has not been pumped for a number of years. The owner of the well would like to know the total thickness of freshwater that might be available for development at the well site and, as a first approximation, has decided to estimate the total thickness by using equations 1 and 2. A topographic map of the area indicates that land surface on the property is about 20 feet above sea level. The depth to the water table measured in the unpumped well is about 10 feet. The elevation of the water table (h) therefore is about 10 feet above sea level, and the estimated thickness of the freshwater zone below sea level (z), estimated by use of equation 2, is 400 feet. The total thickness of freshwater at the site (h + z) therefore is estimated to be about 410 feet.*

Patterns of Freshwater-Saltwater Interaction in a Single-Layer Aquifer, Cape Cod, Massachusetts

Detailed studies of the freshwater-saltwater transition zone in the ground-water flow system of Cape Cod, Massachusetts, illustrate how the configuration of freshwater and saltwater zones can be affected by the distribution of geologic materials within an aquifer and by the distribution of coastal marshes and embayments that bound an aquifer. The Cape's ground-water flow system consists of a single-layer aquifer that is composed of extensive layers of glacially derived sand and gravel interbedded in places with layers and lenses of clay, silt, and unsorted till (LeBlanc and others, 1986). These unconsolidated sediments range in thickness from about 100 ft at Cape Cod Canal to more than 1,000 ft in the town of Truro. The ground-water flow system consists of six separate water-table mounds (fig. 11); ground water moves radially outward from these mounds toward the ocean, saltwater bays, tidal inlets, and coastal wetlands and streams.

Chloride-concentration profiles for 29 monitoring sites distributed across the Cape indicated that the transition between freshwater and saltwater was relatively thin, ranging from about 20 to 80 ft in thickness (LeBlanc and others, 1986). At many of the monitoring sites, such as site BHW 221 (fig. 11), the location and shape of the transition zone was similar to that typically found in relatively thin and homogeneous aquifers; that is, a single transition zone located close to the coastline separated overlying freshwater from underlying saltwater. However, at a number of monitoring sites along Cape Cod Bay, two transition zones have been formed by the presence of very fine sand, silt, and clay deposits (see site A1W 318). At these sites, a shallow transition zone lies above the fine-grained sediments, and fresh ground water discharges directly to the bay; a lower transition zone lies beneath the fine-grained sediments and is displaced offshore. Fresh ground water in the deeper part of the aquifer flows below the fine-grained sediments and then discharges upward into overlying saltwater. At site BMW 47, which is in a saltwater marsh, brackish water having a chloride concentration of about 10,000 mg/L overlies freshwater in the aquifer. The source of the shallow brackish water is periodic flooding by seawater from the bay.

Water-quality data collected at the 29 monitoring sites were used with other hydrogeologic information to develop a conceptual understanding of ground-water flow throughout the Cape Cod aquifer (fig. 12). In the central part of Cape Cod (fig. 12A), the freshwater flow system is truncated by bedrock and fine-grained sediments, whereas on the eastern, outer limb of Cape Cod, the thick deposits of unconsolidated sediments and the generally narrower width of the peninsula cause the freshwater flow system to be underlain everywhere by saltwater (fig. 12B).

Cape Cod National Seashore.

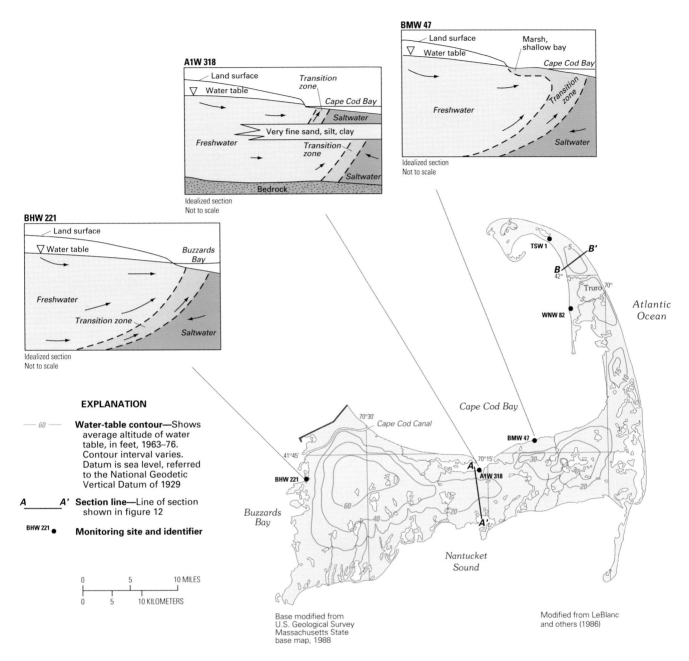

Figure 11. Patterns of the freshwater-saltwater transition zone at three sites in the Cape Cod aquifer, Massachusetts.

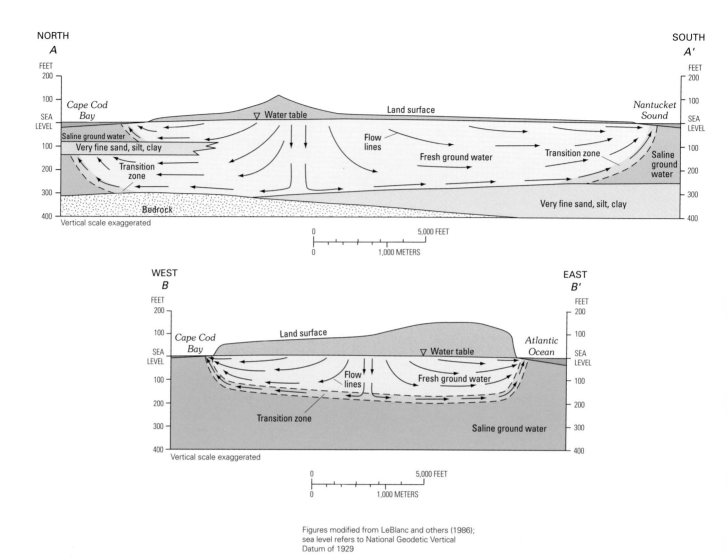

Figure 12. Hydrogeologic sections showing ground-water flow and position of the freshwater-saltwater transition zone, Cape Cod aquifer, Massachusetts. (A) Central Cape Cod; (B) Outer Cape Cod. See figure 11 for locations of sections.

Effects of Sea-Level Fluctuations on the Freshwater-Saltwater Transition Zone, Northern Atlantic Coastal Plain Aquifer System

The Northern Atlantic Coastal Plain encompasses a land area of about 50,000 square miles (mi^2) extending from Long Island, New York, southward to the North Carolina-South Carolina border (fig. 13). The Coastal Plain is underlain by a seaward-thickening wedge of predominantly unconsolidated sediments that increases in thickness from the Fall Line, which is the inland limit of the Coastal Plain, eastward toward the Atlantic Ocean. The Fall Line is so named because of the prevalence of falls and rapids in streams that cross the contact between the hard rocks of the Piedmont Plateau to the west and the less-resistive sediments of the Coastal Plain. The sediment wedge reaches a maximum onshore thickness of about 10,000 ft at Cape Hatteras, North Carolina, but exceeds 7.5 miles in thickness offshore from New Jersey and the Delmarva Peninsula (Trapp and Meisler, 1992). The sediments are mostly gravel, sand, silt, and clay, and have been subdivided into an aquifer system that consists of a vertical sequence of highly permeable aquifers separated by less permeable confining units. Ground-water withdrawals from the aquifer system total more than a billion gallons per day, making it one of the most productive aquifer systems in the United States.

Saltwater underlies freshwater in eastern parts of the regional aquifer system. The transition zone between freshwater and saltwater was delineated throughout the aquifer system in the early 1980s by using geochemical and geophysical data collected at more than 500 locations (Meisler, 1989). The transition zone was defined as the zone of water with chloride concentrations from 250 mg/L to 18,000 mg/L. Generally, chloride concentrations increase in the seaward direction of each aquifer and with depth from the shallowest to the deepest aquifers. Waters within the transition zone probably were produced by the mixing of fresh ground water with either seawater or highly concentrated brines. In the area from Virginia to New Jersey, some of the water samples showed the presence of chloride at concentrations greater than those in seawater, suggesting that the transition zone in that area is largely a mixture of freshwater with a brine. For example, the chloride-concentration profile at well Virginia 57 (fig. 13) shows a maximum chloride concentration of nearly 27,000 mg/L, which is about 8,000 mg/L greater than that of seawater. The most likely source of the brines appears to be the leaching of ancient evaporite deposits of probable early Jurassic age beneath the Continental Shelf and Slope (Meisler, 1989; Knobel and others, 1998).

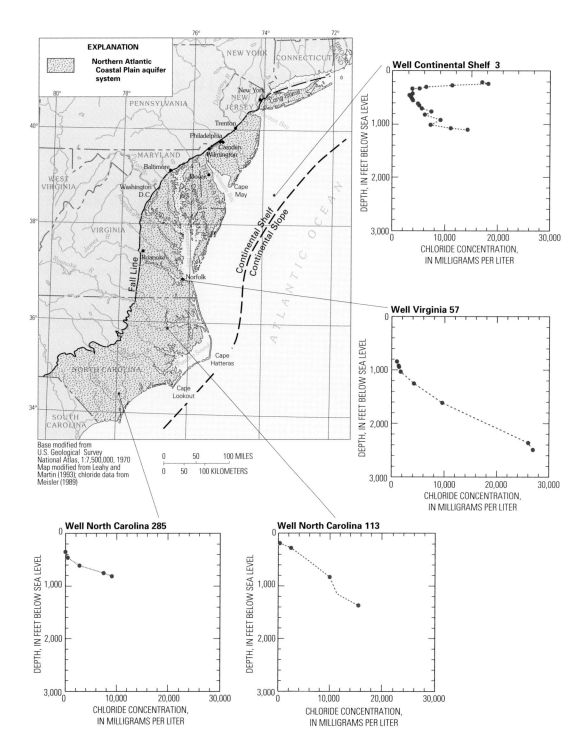

Figure 13. Chloride-concentration profiles at four well sites in the Northern Atlantic Coastal Plain aquifer system. At most sites along the Coastal Plain, there is a transition from freshwater at the top of the aquifer system to saltwater in the deeper parts of the system. This is shown by the chloride-concentration profiles at wells Virginia 57 and North Carolina 113 and 285. However, chloride data from the offshore well site (Continental Shelf 3) indicate a zone of saltwater beneath the seabed, through which freshwater discharges. The reversed transition zone at the site, where saltwater overlies freshwater, probably results from short-term cyclic fluctuations of sea level caused by storms and tides, which temporarily reverse the direction of ground-water flow at the ocean floor and move seawater into the aquifer (Meisler and others, 1985).

Two striking features of the transition zone within the regional aquifer system are its large vertical thickness and substantial horizontal width. The thickness of the transition zone ranged from 400 to 2,200 ft, whereas the width of the transition zone was as much as 40 miles in some areas (fig. 14). The development of the broad transition zone has been attributed to the cyclic movement of saltwater caused by global sea-level fluctuations that resulted in repeated advance and retreat of the freshwater-saltwater interface during at least the last 900,000 years (Meisler and others, 1985). As the sea level rose, saltwater invaded the aquifer sediments and mixed with freshwater. As the sea level declined, the fresher water advanced seaward, and the process of mixing continued. Repeated advance and retreat of the saltwater produced a broad zone of mixed waters in which saltwater predominates in the deeper and seaward parts, and freshwater predominates in the shallower and landward parts (Meisler and others, 1985).

The depth to the top of the transition zone is shallowest in North Carolina and deepens northward, reaching its greatest depths—as much as 2,800 ft below sea level—in Maryland and along the coast of New Jersey. Moreover, ground water containing chloride concentrations of less than 5,000 mg/L has been found as much as 55 mi from the New Jersey coast, but extends progressively shorter distances from the coast southward to Virginia and North Carolina (Meisler, 1989). The occurrence of the transition zone at great depths in New Jersey and Maryland and the occurrence offshore of water considerably fresher than seawater have been attributed to long periods when sea levels generally were lower than at present (Meisler, 1989). Overall, the average sea level during the past 900,000 years is estimated to have been about 150 ft lower than the present sea level. It has been hypothesized that, at least in some areas, the transition zone may not be in equilibrium with the present-day sea level, but may still be moving landward and upward to adjust to the present sea level (Meisler and others, 1985; Pope and Gordon, 1999).

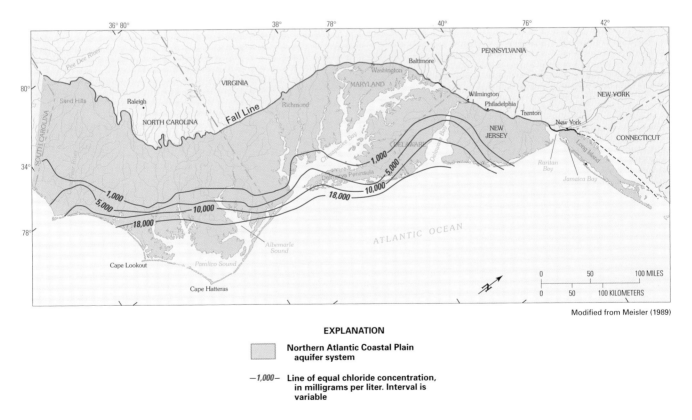

Figure 14. Chloride concentrations at the top of the middle Potomac aquifer, Northern Atlantic Coastal Plain aquifer system.

Virginia's Inland Saltwater Wedge and the Chesapeake Bay Impact Crater—The Importance of Geologic Control on Coastal Ground-Water Systems

For many years, scientists have been puzzled by an anomalous wedge of saltwater that extends inland from the mouth of Chesapeake Bay into the Coastal Plain aquifers beneath the southeastern parts of the Middle and York-James Peninsulas of Virginia. This inland wedge is shown on regional maps of the freshwater-saltwater transition zone in the Northern Atlantic Coastal Plain aquifers and, in greater detail, on a map of dissolved-solids concentrations for the upper Potomac aquifer of Virginia (fig. 15). Several hypotheses have been proposed to explain the occurrence of this saltwater wedge. As early as 1911, scientists suggested that the wedge was the product of incomplete flushing of ancient seawater that had invaded the aquifers from above during high stands of the sea. The incomplete flushing of the saltwater was attributed to poor circulation of ground water caused by a decrease in the permeability or by a pinching out of the aquifers in the area of the wedge. Others hypothesized that the wedge was the result of relatively low ground-water levels in the area of major ground-water discharge to lower Chesapeake Bay. Scientists also have noted that chloride concentrations in the wedge often are greater than those in seawater, which has led to speculation that the source of the saltwater was a brine.

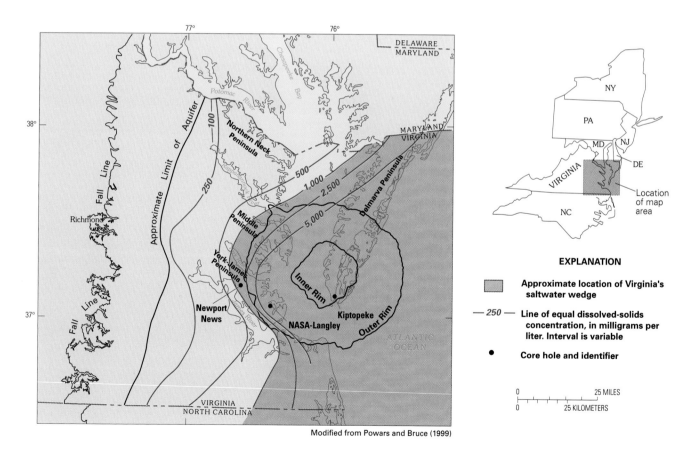

Figure 15. Relation of dissolved-solids concentrations in the upper Potomac aquifer to the location of the inner and outer rims of the Chesapeake Bay impact crater.

During the early 1990s, scientists studying the geologic structure and history of the Atlantic Coastal Plain made an important discovery that is dramatically altering the understanding of Virginia's inland saltwater wedge. Through the collection and analysis of deep sedimentary cores, geophysical data, and fossil assemblages, geologists have identified a large impact crater formed by a meteorite near what is now the mouth of Chesapeake Bay (figs. 16 and 17) (Poag and others, 1994; Poag, 1998, 1999, and 2000; Powars and Bruce, 1999; Powars, 2000). The crater is three times larger than any other U.S. crater and is the sixth largest crater known on Earth. Discovery of the crater helps to explain several of the previously anomalous geologic and hydrologic conditions in southeastern Virginia and is an excellent example of the important role of geologic control on ground-water-flow systems and ground-water quality.

Approximately 35 million years ago, a large meteorite struck the shallow waters of the Continental Shelf of what is now southeastern Virginia, blasting through the water column and thousands of feet of unconsolidated sediment and underlying granitic crust. Sea level was high everywhere on Earth at the time of the impact, and the ancient shoreline in the Virginia region was somewhere in the vicinity of present-day Richmond (fig. 16). The meteorite measured 1 to 2 miles in diameter and was traveling about 50,000 miles per hour (Poag, 2000). The impact likely vaporized seawater and the crystalline granitic bedrock immediately beneath the impact site and created a supersonic shock wave, tsunamis (great sea waves traveling at very high speeds), and a huge atmospheric vapor cloud. The impact produced an enormous, 56-mile-wide crater that reaches about 6,000 ft deep—a depth comparable to the Grand Canyon of the Colorado River. The crater was immediately filled with a chaotically mixed, sandy rubble bed known as breccia (fig. 17). The breccia contains hand-size to person-size chunks (clasts) of clay, limestone, and sand (Poag, 1998, 1999). These clasts are fragments of the geologic beds that were removed by the impact and small pieces of the underlying granitic basement rocks. Along the crater's outer rim, large-scale blocks of sediments (megaslump blocks) collapsed into the crater. The entire impact event, from initial atmospheric penetration to the end of breccia deposition, lasted only a few hours or days. The crater then was buried by additional sedimentary beds that accumulated during the following 35 million years.

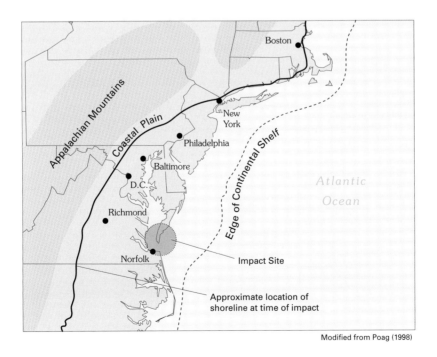

Figure 16. Location of the Chesapeake Bay impact crater and of the shoreline at the time of impact.

The Chesapeake Bay impact crater is the most significant discontinuity in the sedimentary strata of the Atlantic Coastal Plain. Its discovery has led to a re-interpretation of the hydrogeologic framework and inland saltwater wedge of southeastern Virginia. Previous views of Coastal Plain sediments in the area were of a layered sequence of alternating aquifers and intervening confining units similar to the structure of aquifers and confining units throughout most of the Northern Atlantic Coastal Plain aquifer system (see, for example, fig. 5). The new interpretation, however, is that these aquifers and confining units were truncated and excavated by the impact, which left in their place rubbly breccias. The poorly sorted breccias are thought to be of lower permeability than the surrounding aquifers and, therefore, likely affect ground-water flow patterns near and within the crater.

The spatial association of the inland saltwater wedge with the configuration and location of the outer rim of the meteorite crater has led scientists to propose that the impact crater may have been the cause, or at least part of the cause, of the saltwater wedge. The association between the saltwater wedge and crater is shown by lines of equal dissolved-solids concentrations drawn for one of the aquifers that abuts the crater (the upper Potomac aquifer; fig. 15). Although it currently (2003) is not known for certain whether the crater is the ultimate cause or one of the causes of the saltwater wedge, it is clear that the impact crater affects the framework of the aquifers and the ground-water conditions in southeastern Virginia. Moreover, geophysical surveys across the crater show many faults that cut the sedimentary beds above the breccia and extend upward toward the floor of Chesapeake Bay (Poag, 1998, 1999). These faults are a result of the compaction subsidence of the

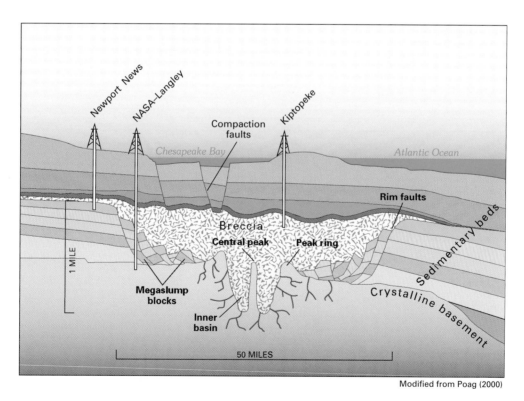

Figure 17. Schematic cross section showing some of the features of the Chesapeake Bay impact crater and location of three core holes that provided data on these features. The impact breccia fills the excavated basement rocks, surrounded by undisturbed sedimentary rocks, and overlain by post-impact sediments. (Locations of core holes shown in figure 15.)

breccia that has occurred since the impact. Some of the faults appear to completely breach the confining unit over the saltwater wedge, which could allow upward flow of saltwater.

Continuing investigations of the geology and hydrology of the impact crater are underway to better understand how the crater affects ground-water resources of the Virginia Coastal Plain. During the summer of 2000, a 2,084-ft-deep core hole was drilled through the crater at the NASA Langely Research Center in Hampton, Virginia (fig. 17). This core hole is the first of several that are planned to be drilled in and around the crater by a multiagency collaboration that was formed to continue scientific investigations of the crater (Gohn and others, 2001). These investigations could provide information on the hydrogeologic framework, ground-water flow system, and water quality in and near the crater. One of the planned uses of this information will be the revision of a computer model of the ground-water system of the Virginia Coastal Plain. This model was first developed in the early 1980s and is used by the State of Virginia to help manage the more than 100 million gallons per day (Mgal/d) of ground water that are withdrawn from Coastal Plain aquifers in the State.

U.S. Geological Survey scientist collects sediment core at a site within the Chesapeake Bay impact crater, Hampton, Virginia, during the summer of 2000 (top). Drilling into the Chesapeake Bay impact crater at NASA Langley Research Center, Hampton, Virginia, during the summer of 2000 (bottom).

Saltwater in the Floridan Aquifer System, South Carolina to Florida

The Floridan aquifer system is one of the most productive aquifers in the world. It consists of a thick sequence of carbonate rocks (limestones and dolomites) that underlie all of Florida, southern Georgia, and small parts of adjoining South Carolina and Alabama; a total area of about 100,000 mi^2 (fig. 18). An estimated 4.0 Bgal/d of water was withdrawn from the aquifer system in 2000 (R.L. Marella, U.S. Geological Survey, written commun., 2003), and, in many areas, it is the sole source of freshwater (Johnston and Bush, 1988). In addition to water supply, the Floridan is being used increasingly for aquifer storage and recovery systems, in which freshwater is injected into more saline zones of the aquifer and stored for later use. Moreover, in several places where the aquifer system contains saltwater, such as along the southeastern coast of Florida, treated sewage water and industrial wastes are injected into it.

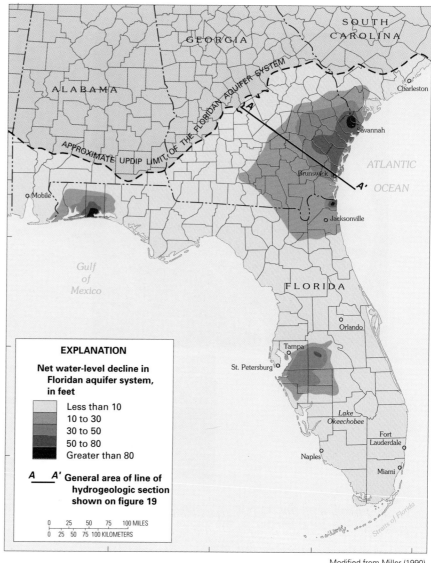

Figure 18. Areas of large, regional water-level declines in the Floridan aquifer system, 1980.

Modified from Miller (1990)

The aquifer system generally thickens seaward from a thin edge near its northern limit to a maximum of about 3,500 ft in southwestern Florida. In most places, the system consists of the Upper and Lower Floridan aquifers separated by a less permeable confining unit (the "middle confining unit") that restricts movement of water between the two aquifers (fig. 19). Much of the aquifer system is overlain by an upper confining unit that, where present, limits the amount of recharge to the system. Where the upper confining unit is thin or absent, recharge is plentiful and ground-water circulation is high. In these areas of high recharge and vigorous circulation, ground water readily dissolves the carbonate rocks that make up the aquifer system, creating large and highly permeable conduits that store and transmit tremendous volumes of ground water. These large conduits are the cause for the many first-magnitude springs—those with a flow of 100 cubic feet per second or more—that issue from the aquifer system.

Ground-water withdrawals have resulted in long-term regional water-level declines of more than 10 ft in three broad areas of the flow system: (1) coastal Georgia and adjacent South Carolina and northeast Florida; (2) west-central Florida; and (3) the Florida panhandle (fig. 18). In these and a number of other coastal areas, ground-water withdrawals have reversed the generally seaward direction of ground-water flow, creating the potential for saltwater intrusion from the Gulf of Mexico or Atlantic Ocean or from deep parts of the aquifer that contain saltwater.

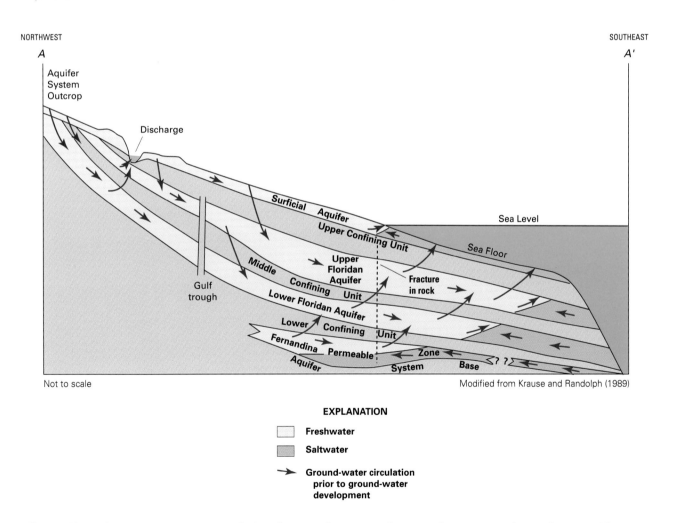

Figure 19. Schematic cross section of Floridan aquifer system from northwest to southeast Georgia. The Lower Floridan aquifer in this area includes a highly permeable unit called the Fernandina permeable zone. This zone is the source of a considerable volume of fresh to brackish water that leaks upward through the middle confining unit and ultimately reaches the Upper Floridan aquifer (Miller, 1990). Line of section shown on figure 18.

The transition between freshwater and saltwater in the Floridan aquifer system is illustrated by the distribution of chloride in water in the Upper and Lower Floridan aquifers (fig. 20). Although large areas of the Upper Floridan aquifer contain water with a chloride concentration less than 250 mg/L (fig. 20A), much of the Lower Floridan aquifer contains water with chloride concentrations that exceed the 250 mg/L drinking-water limit (fig. 20B), which has limited the aquifer's use for water supply. In general, chloride concentrations in the Upper Floridan aquifer are related to ground-water-flow conditions and proximity to the coast. In areas where the upper confining unit is thin or absent, fresh ground-water circulation rates are high and chloride concentrations tend to be low (less than 250 mg/L). Where the flow system is tightly confined, flow is more sluggish and

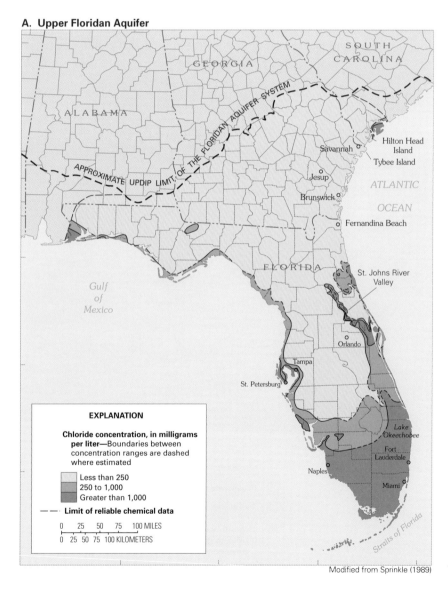

Figure 20. Chloride concentrations in water from the upper 200 feet of the Upper Floridan aquifer (A, above) and from the Lower Floridan aquifer (B, facing page). A number of localized areas of high chloride concentrations caused by saltwater intrusion in response to ground-water withdrawals, such as in Brunswick, Georgia, are not shown.

chloride concentrations in the aquifer are higher. This is the case in Florida south of Lake Okeechobee (fig. 20A), where the Upper Floridan aquifer is extensively confined and ground-water flow is quite sluggish. Because of the slow movement of ground water in the area, it is thought that residual seawater that entered the aquifer during the Pleistocene when sea level was higher than its current level has not been completely flushed out by modern freshwater (Sprinkle, 1989; Johnston and Bush, 1988; Reese, 1994, 2000; Reese and Memberg, 2000). The anomalously high concentrations of chloride along the St. Johns River and the eastern coast of Florida are thought to be the result, in varying amounts, of two processes (Sprinkle, 1989): (1) incomplete flushing by the modern-day freshwater flow system of residual seawater that invaded the aquifer during high sea-level stands

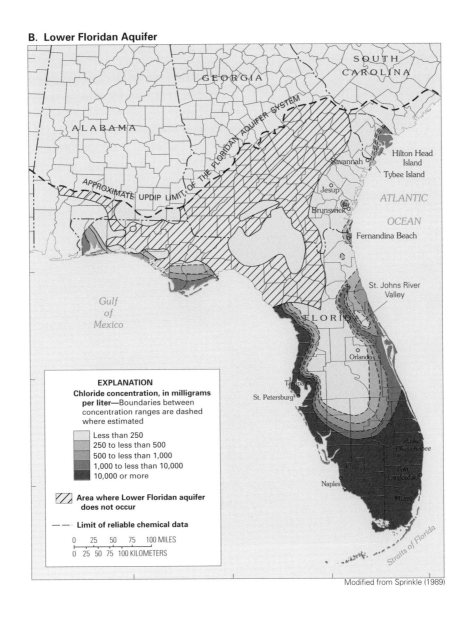

Modified from Sprinkle (1989)

of the Pleistocene; and (2) upward flow of brackish water from the underlying Lower Floridan aquifer along fracture zones in the aquifer system. The high chloride concentrations in the western panhandle of Florida also may have resulted from incomplete flushing of residual seawater from Pleistocene highstands.

The generally low chloride concentrations in the Upper Floridan aquifer along the Georgia coast (fig. 20A) have been attributed to the thick confining unit that overlies the Upper Floridan aquifer in that area. The confining unit has created relatively high ground-water heads that have kept the freshwater-saltwater interface offshore. In fact, freshwater flow has been observed to extend as far as 50 mi offshore of southeast Georgia (Johnston and others, 1982). However, along the coast in South Carolina and extending to Tybee Island, Georgia, high chloride concentrations in water from the aquifer system are attributed to intrusion of offshore saltwater caused by large ground-water withdrawals from the Upper Floridan aquifer in the Savannah, Georgia, and the Hilton Head Island, South Carolina, areas. Saltwater most likely enters the aquifer system by lateral intrusion from offshore areas combined with some downward vertical leakage of seawater to the Upper Floridan aquifer where the overlying confining unit is thin or absent (Krause and Clarke, 2001).

Ground-water temperature and geochemical data from the Lower Floridan aquifer in the south Florida area suggest that a geothermally driven circulation of ground water occurs in the Lower Floridan aquifer in that area (Kohout, 1965; Meyer, 1989; Sanford and others, 1998). In this circulation, cold, dense ocean water enters the Lower Floridan aquifer where it is in direct contact with the ocean. As the seawater moves landward, it is warmed by geothermal heat generated naturally below the base of the thick Floridan aquifer system that underlies the Florida peninsula. The geothermal heating lowers the density of the seawater, causing it to rise where it is diluted and transported back to the ocean with seaward-flowing water of lower salinity.

In 1979, an oil exploration well located about 55 miles offshore of St Marys, southeast Georgia, detected freshwater in the upper part of the Upper Floridan aquifer (Johnston and others, 1982).

CHAPTER 2. CAUSES, MODES, AND MANAGEMENT OF SALTWATER INTRUSION

The natural balance between freshwater and saltwater in coastal aquifers is disturbed by ground-water withdrawals and other human activities that lower ground-water levels, reduce fresh ground-water flow to coastal waters, and ultimately cause saltwater to intrude coastal aquifers. Although ground-water pumping is the primary cause of saltwater intrusion along the Atlantic coast, lowering of the water table by drainage canals has led to saltwater intrusion in a few locations, notably southeastern Florida. Other hydraulic stresses that reduce freshwater flow in coastal aquifers, such as lowered rates of ground-water recharge in sewered or urbanized areas, also could lead to saltwater intrusion, but the impact of such stresses on saltwater intrusion, at least currently (2003), likely is small in comparison to pumping and land drainage.

The variability of hydrogeologic settings, sources of saline water, and history of ground-water withdrawals and freshwater drainage along the Atlantic coast has resulted in a variety of modes of saltwater intrusion across the region. Saltwater can contaminate a freshwater aquifer through several pathways, including lateral intrusion from the ocean; by upward intrusion from deeper, more saline zones of a ground-water system; and by downward intrusion from coastal waters. A few of these pathways are illustrated on figures 21 and 22. Some authors have used the term *saltwater encroachment* to refer to lateral movement of saltwater within an aquifer and the term *saltwater intrusion* to refer to vertical movement of saltwater. This distinction is not made in this report because there are often both lateral and vertical components to a particular saltwater-intrusion problem, and because the net result—contamination of a freshwater aquifer—is the same for either type of saltwater movement. Another term that has been used to describe a specific type of vertical saltwater intrusion is *saltwater upconing*, which refers to the movement of saltwater from a deeper saltwater zone upward into the freshwater zone in response to pumping at a well or well field (Reilly and Goodman, 1987). An example of saltwater upconing that occurred on Cape Cod, Massachusetts, is shown in figure 22.

Saltwater intrusion reduces freshwater storage in coastal aquifers and can result in the abandonment of freshwater supply wells when concentrations of dissolved ions exceed drinking-water standards. Saltwater intrusion has been documented throughout the Atlantic coastal zone for more than 100 years (Barlow and Wild, 2002), but the degree of saltwater intrusion varies widely among localities and hydrogeologic settings. In many instances, the area contaminated by saltwater is limited to small parts of the aquifer

Irrigation well pumping from the Floridan aquifer system near Tampa, Florida. Large ground-water withdrawals from the Floridan aquifer system have caused water-level declines and saltwater intrusion in some areas of South Carolina, Georgia, and Florida.

Photograph courtesy of Richard Marella, U.S. Geological Survey

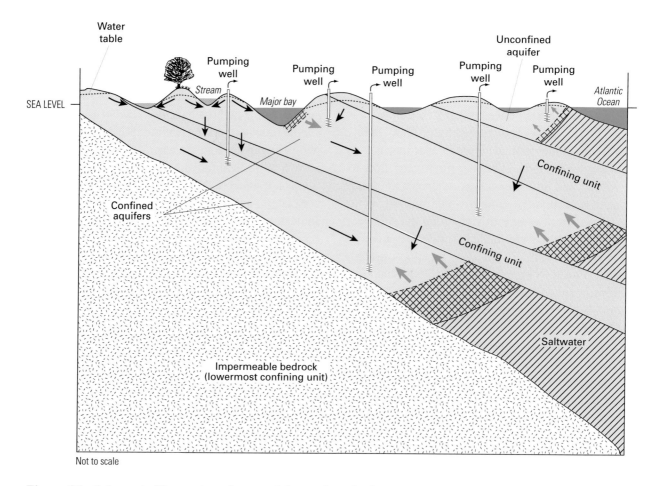

Figure 21. *Schematic illustration of some of the modes of saltwater intrusion in a multilayer, regional aquifer system caused by ground-water pumping at wells. Saltwater moves into the unconfined aquifer from the Atlantic Ocean and into the shallow part of the top confined aquifer from the major bay. The two freshwater-saltwater interfaces at the seaward boundary of each of the confined aquifers also move landward as saltwater is drawn inland from offshore areas.*

and has little or no effect on wells pumped for ground-water supply (Box C). In other instances, contamination is of regional extent and has substantially affected ground-water supplies. For example, in Cape May County, New Jersey, more than 120 supply wells have been abandoned since 1940 because of saltwater contamination (Lacombe and Carleton, 1992). The extent of saltwater intrusion into an aquifer depends on several factors, including the total rate of ground water that is withdrawn compared to the total freshwater recharge to the aquifer, the distance of the stresses (wells and drainage canals) from the source (or sources) of saltwater, the geologic structure and distribution of hydraulic properties of the aquifer, and the presence of confining units that may prevent saltwater from moving vertically toward or within the aquifer. Moreover, the time required for saltwater to move through an aquifer and reach a pumping well can be quite long. Depending on the location and lateral width of the transition zone, many years may pass before a well that is unaffected by saltwater intrusion suddenly may become contaminated.

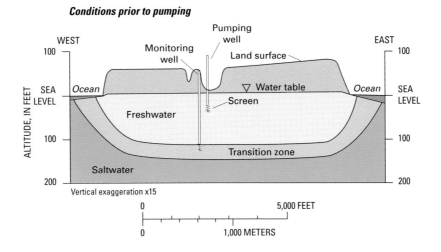

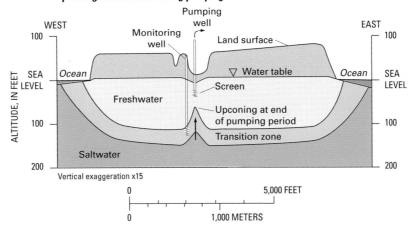

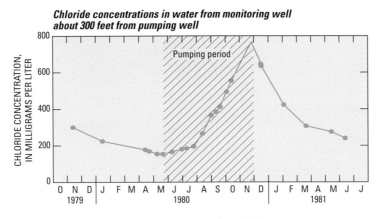

Figures modified from LeBlanc and others (1986)

Figure 22. Ground-water pumping from a well in the town of Truro on Cape Cod, Massachusetts, caused upconing of the transition zone beneath the well, which in turn caused increased chloride concentrations in a nearby monitoring well. After pumping at the well stopped in November 1980, the transition zone moved downward, and chloride concentrations at the monitoring well slowly decreased with time.

Saltwater Intrusion in a Fractured Crystalline-Rock Aquifer, Harpswell, Maine

Large parts of the northeastern Atlantic coastal zone are underlain by crystalline igneous and metamorphic rocks (often referred to as bedrock) that generally yield only small amounts of water to wells. However, because these rocks extend over large areas, large volumes of ground water are withdrawn from them, and in many places, they are the only reliable water sources (Miller, 1999). Crystalline-rock aquifers are geologically complex systems in which the movement of ground water is strongly dependent on the presence of fractures, joints, faults, and other openings in the rock mass. The potential for saltwater intrusion into these aquifers exists where the openings extend to the coast or to depths at which they become saturated with saltwater. Although incidences of saltwater intrusion into crystalline-rock aquifers of the northeastern United States have been reported, the problem does not appear to be widespread. In a review of the coastal bedrock aquifers of Maine, Caswell (1979a, b) noted that most coastal wells do not pump saltwater, even though the ocean nearly surrounds the numerous islands and narrow peninsulas of Maine's irregular coastline. This may be partly the result of the generally low pumping rates of wells that tap the bedrock aquifers, but Caswell suggested that the absence of extensive saltwater intrusion is probably the result of the particular orientation of the water-bearing fractures of crystalline rocks along the coastline, which prevents, or at least limits, the landward migration of saltwater. For example, the principal water-bearing fractures of the bedrock aquifer on High Head peninsula in Harpswell, Maine (fig. C–1), are nearly vertical and are parallel to the length of the peninsula and to the coastline. Of more than 60 bedrock wells that had been drilled on the peninsula as of the mid-1970s, only 3 had been contaminated by saltwater (Richard, 1976; Caswell, 1979a, b). These three wells were in an area where the principal water-bearing fractures have been cut by a number of large fractures that lie perpendicular to the length of the peninsula and the dominant fracture direction (fig. C–1). These cross-fractures appear to make the rock very permeable perpendicular to the coastline, and to provide a conduit for saltwater to move readily toward the wells.

Fractures and other openings in crystalline rocks such as these along the rocky coast of Acadia National Park, Maine, can provide conduits for saltwater to enter coastal aquifers.

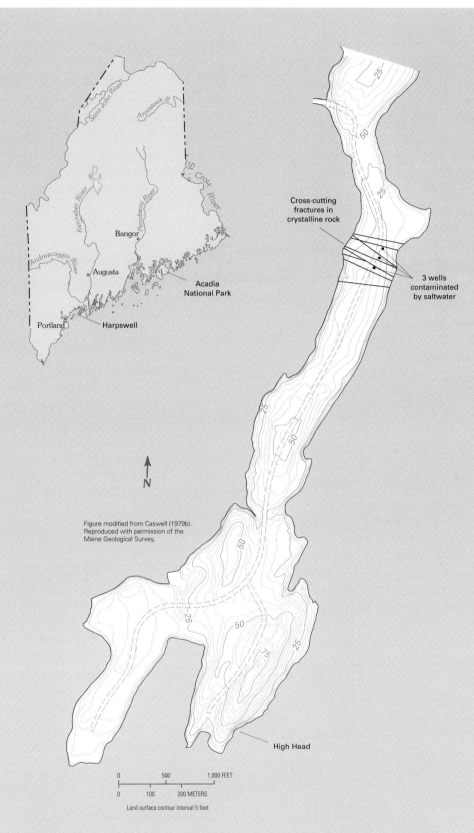

Figure C–1. *High Head peninsula, Harpswell, Maine, where saltwater contamination of three domestic wells in a fractured crystalline-rock aquifer was documented in the mid-1970s.*

Many States and communities along the Atlantic coast are taking actions to manage and prevent saltwater intrusion to ensure a sustainable source of ground water for the future. These actions can be grouped broadly into three general categories: engineering techniques, regulatory (legislative) approaches, and scientific monitoring and assessment (van Dam, 1999). Often, several actions are taken simultaneously or as part of a more comprehensive strategy for managing both the ground-water and surface-water supplies of a coastal community. The State of Georgia, for example, has established a multicomponent approach for managing saltwater intrusion that restricts withdrawals in some coastal areas, encourages water conservation, and relies on hydrologic studies and water-quality monitoring to better understand saltwater movement in the State's aquifers and to evaluate alternative sources of freshwater. The regional scale of many of the aquifers along the Atlantic coast means that several communities and political jurisdictions often share a single aquifer or aquifer system, and, therefore, ground-water development in one community can affect the water resources of neighboring communities.

A common approach for managing saltwater intrusion has been to reduce the rate of pumping from coastal wells or to move the locations of withdrawals further inland. Reductions in coastal withdrawals allow ground-water levels to recover from their lowered (or stressed) levels, and fresh ground water to displace the intruded saltwater. In New Jersey, for example, State-mandated reductions in ground-water withdrawals in some coastal counties have resulted in ground-water-level increases in aquifers that have been affected by saltwater intrusion (Box D). An alternative to reducing ground-water withdrawals is to artificially recharge freshwater into an aquifer to increase ground-water levels and hydraulically control the movement of the intruding saltwater. Artificial recharge can be accomplished through injection wells or by infiltration of freshwater at the land surface. In either case, the recharged water creates hydraulic barriers to saltwater intrusion.

Perhaps the most prominent example of the use of artificial recharge to control saltwater intrusion on the east coast of the United States is in southeastern Florida. In that area, an extensive network of surface-water canals is used to convey freshwater from inland water-conservation (storage) areas during the dry season to coastal areas, where the water is recharged through the canals to the underlying Biscayne aquifer to slow saltwater intrusion in the aquifer.

In addition to more conventional methods, innovative approaches are now being used to manage saltwater intrusion along the Atlantic coast. These include aquifer storage and recovery systems, desalination systems, and blending of waters of different quality. Aquifer storage and recovery (ASR) is a process by which water is recharged through wells into a suitable aquifer, stored for a period of time, and then extracted from the same wells when needed (Pyne, 1995). Typically, water is stored during wet seasons and extracted during dry seasons. ASR systems have been developed in Wildwood (Cape May County), New Jersey, the town of Chesapeake, Virginia, and at several locations in Florida.

Desalination refers to water-treatment processes that produce freshwater by removing dissolved salts from brackish or saline waters (Buros, 2000). Desalination systems are becoming more widespread in the United States as desalination technologies improve, costs decrease, and new sources of freshwater become more difficult to develop. One of the interesting aspects of the increased use of desalination systems is that it changes the perspective on brackish or saline water from that of a potential water problem (a contaminant) to that of a potential water source. Past studies have highlighted the importance of saline ground water as a potential resource (Kohout, 1970) and estimated the saline ground-water resources of the conterminous United States (Feth and others, 1965). Despite these efforts, substantial research remains to be done to characterize the distribution and quality of brackish and saline ground waters that might be used for desalination systems (Alley, 2003).

Scientific monitoring and assessment provide basic characterization of the ground-water resources of an area, an understanding of the different pathways by which saltwater may intrude an aquifer, and a basis for management of water supplies. Numerous water-level and water-quality monitoring networks have been established along the Atlantic coast to monitor changes in ground-water levels, ground-water quality, and movement of the freshwater-saltwater interface. Water-quality monitoring networks are particularly important in serving as early-warning systems of saltwater movement toward freshwater supply wells; their use along the Atlantic coast is described in greater detail in the following chapter.

Six case studies are provided in the remainder of this chapter to illustrate specific causes and modes of saltwater intrusion along the Atlantic coast and to describe several approaches that have been used to manage and prevent saltwater intrusion in this coastal region.

The six examples are from the States of New Jersey, Georgia, and Florida, which, with the possible exception of Long Island, New York (fig. 23), have experienced the most widespread problems of saltwater intrusion along the Atlantic coast. The case studies demonstrate several factors that must be considered in the analysis and management of a saltwater-intrusion problem, specifically: the source (or sources) of saltwater to the aquifer, the physical structure and dynamics of the aquifer or aquifer system into which the saltwater intrudes, and the types and locations of stresses to the aquifer (supply wells and drainage canals). The first two case studies are drawn from areas within the Northern Atlantic Coastal Plain aquifer system, which consists of unconsolidated to semiconsolidated sand and gravels. The last four case studies describe examples of saltwater intrusion into carbonate aquifers and aquifer systems of the southeastern States.

Photographs courtesy of Dare County Water Department, Kill Devil Hills, North Carolina

The North Reverse Osmosis (RO) Water Plant is part of the water-supply system for Dare County, which is on the Outer Banks of North Carolina. The RO system has been in operation since August 1989 and at the time of its initiation was the largest RO system outside the State of Florida. The plant is supplied with brackish ground water that is pumped from the Yorktown aquifer by 10 wells at the facility, each of which can produce as much as 500 gallons per minute (Robert W. Oreskovich, Dare County Water Department, Kill Devil Hills, North Carolina, June 2003). The treatment system (bottom photograph) consists of three reverse osmosis units, each of which produces 850,000 gallons per day of treated water. The treated water, which is referred to as the permeate, then is mixed with 150,000 gallons per day of raw brackish ground water, for a total production of 1.0 million gallons per day of potable water for each RO unit.

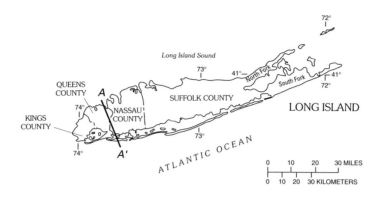

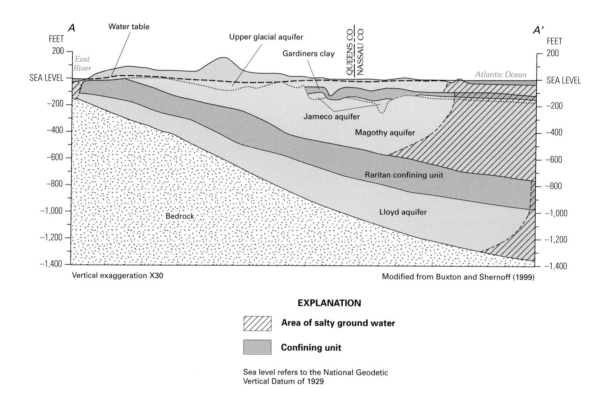

Figure 23. Saltwater occurrence and intrusion in the aquifers of Long Island, New York. Freshwater on the island discharges along most of the periphery of the island, which prevents saltwater from entering the aquifers. In western Long Island, however, saltwater wedges that are hydraulically connected to the sea are found in aquifers on the Atlantic Ocean side of the island. The saltwater wedge in the Lloyd aquifer extends seaward of the wedges in the overlying aquifers because of the relatively impermeable clays of the Raritan confining unit, which force freshwater in the Lloyd aquifer to extend seaward of the island. The positions of the saltwater wedges have been attributed mainly to natural conditions that prevailed long before the start of ground-water development in western Long Island. Ground-water pumping, however, has caused a landward migration of the freshwater-saltwater interface in aquifers in western Long Island since the late 1890s (Lusczynski and Swarzenski, 1966; Buxton and Shernoff, 1999). Saline ground water probably also is migrating downward into the Lloyd aquifer from the overlying Jameco and Magothy aquifers in areas of heavy pumping. In the Forks areas of eastern Long Island, saltwater underlies freshwater in lens-shaped reservoirs that resemble those that underlie outer areas of Cape Cod, Massachusetts (see figure 12B) (Nemickas and Koszalka, 1982).

Water-Level Response to Mandated Decreased Withdrawals in the New Jersey Coastal Plain

Ground-water withdrawals from the 10 major confined aquifers in the New Jersey Coastal Plain (fig. D–1), which is part of the Northern Atlantic Coastal Plain, began in the late 1800s and were approximately 210 Mgal/d in 1997 (Lacombe and Rosman, 2001). These withdrawals have created several regions of large ground-water-level declines, with declines of more than 200 ft below sea level in some areas. Since 1978, the USGS, in cooperation with the New Jersey Department of Environmental Protection (NJDEP), has measured ground-water levels every 5 years in more than 700 monitoring wells to provide a "snapshot" of ground-water conditions in each of the confined aquifers. In 1985–86, the NJDEP used these water-level measurements to establish two "Water Supply Critical Areas" in the State—areas where excessive water usage threatens the long-term sustainability of the water-supply source.

Water levels in the four aquifers underlying Critical Area 1 in parts of Middlesex, Monmouth, and Ocean Counties—specifically, the Wenonah-Mount Laurel (fig. D–2), Englishtown, and the Upper and Middle Potomac-Raritan-Magothy aquifers—have been lowered below sea level as a result of ground-water withdrawals, causing saltwater to flow landward toward the aquifers. In response to these lowered ground-water levels, NJDEP mandated that withdrawals from aquifers underlying the Critical Area be reduced after 1988, and that surface water from reservoirs located in Monmouth County be used as a substitute. Upon completion of the Manasquan Reservoir in 1991, water users in Critical Area 1 began to withdraw less water from the confined aquifers.

As a result of the mandated reduced withdrawals, ground-water levels in aquifers underlying Critical Area 1 rose by as much as 120 ft between the 1988 and 1993 water-level measurements, and there was a reduction in the overall breadth of the area of lowered ground-water levels (Lacombe and Rosman, 1997). For example, water levels in observation well 25–486 in eastern Monmouth County rose from about 180 ft below sea level to about 90 ft below sea level during the 5-year period (fig. D–3). Additional water-level increases of as much as 47 ft were measured in the aquifer between the 1993 and 1998 water-level measurements (Lacombe and Rosman, 2001).

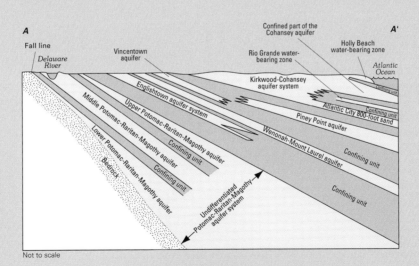

Figure D–1. Diagrammatic hydrogeologic section of the New Jersey Coastal Plain. Line of section shown on figure D–2.

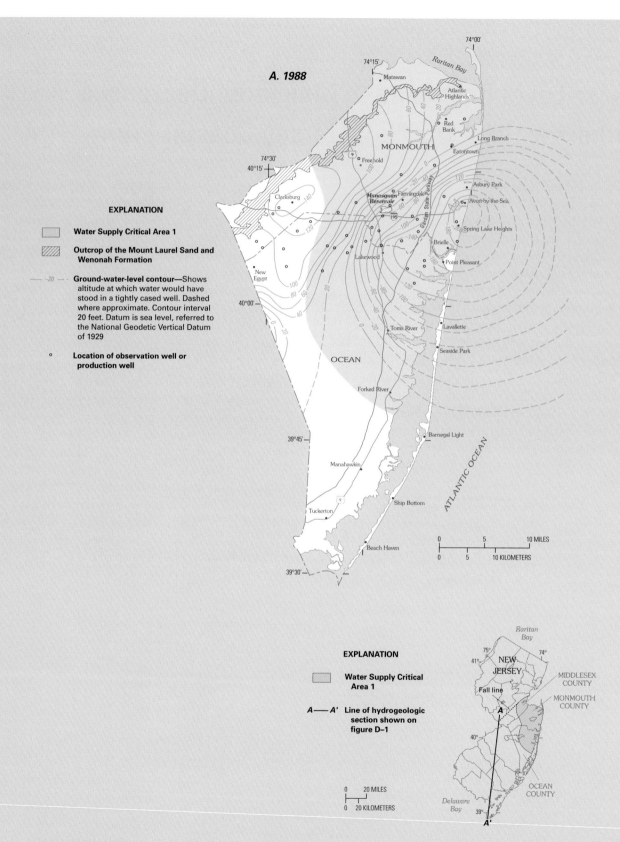

Figure D–2. Ground-water-level contours for the Wenonah-Mount Laurel aquifer, (A) 1988 and (B) 1993.

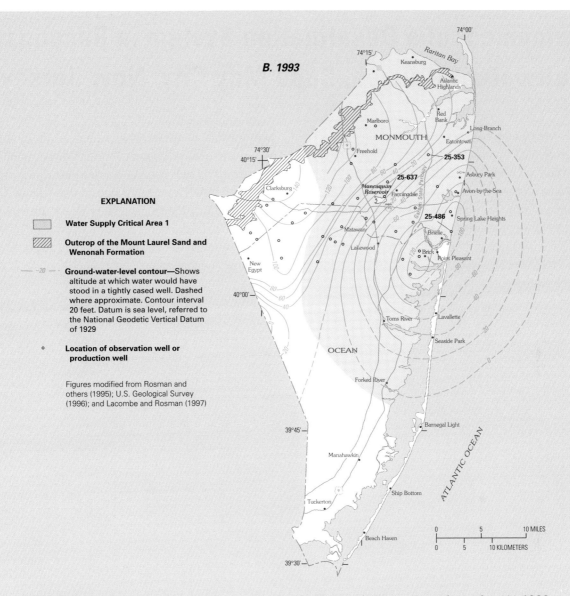

Figure D–2. Ground-water-level contours for the Wenonah-Mount Laurel aquifer, (A) 1988 and (B) 1993—Continued.

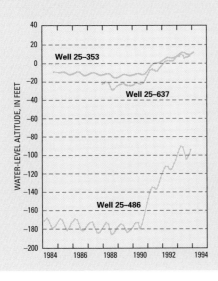

Figure D–3. Water-level hydrographs for observation wells screened in the Wenonah-Mount Laurel aquifer, 1984–94. Seasonal water-level fluctuations of 5 to 15 feet occurred in well 25–486 in response to seasonal fluctuations in ground-water withdrawal rates.

Development of a Desalination System in Response to Saltwater Intrusion, Cape May City, New Jersey

Cape May County, New Jersey, is on a peninsula that is surrounded by the salty waters of Delaware Bay and the Atlantic Ocean (fig. 24). The county is a popular vacation area that has experienced substantial increases in resident and summer tourist populations. Withdrawals from five freshwater aquifers that underlie the peninsula (fig. 25) have lowered ground-water levels as much as 100 ft in some areas and have caused saltwater to intrude each of the aquifers. Prior to the start of ground-water withdrawals, water levels in each of the aquifers stood above sea level, and ground water flowed radially outward from inland recharge areas to low-lying streams, tidal wetlands, Delaware Bay, and the Atlantic Ocean. Saltwater intrusion in the county began about 1890 after the first deep wells were pumped and fresh ground-water levels declined to below sea level for the first time (Lacombe and Carleton, 1992, 2002; Spitz, 1998).

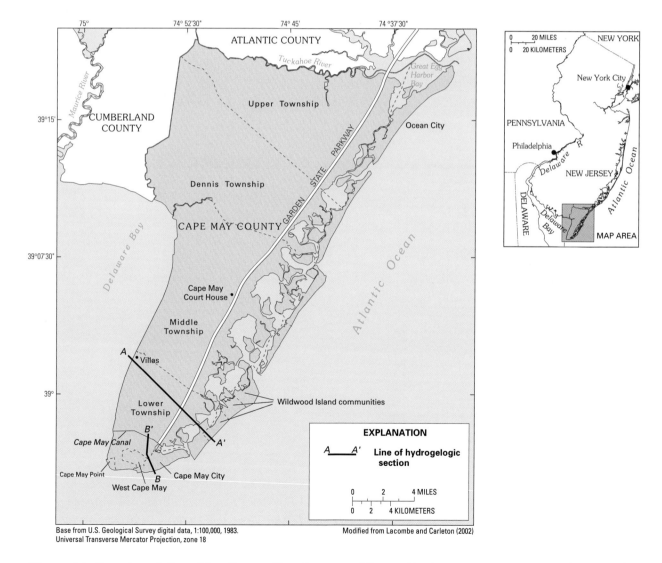

Figure 24. Location of Cape May County, New Jersey, and lines of hydrogeologic sections A-A' and B-B'.

The area affected by saltwater intrusion is along the populated coast of southern Cape May peninsula, including Cape May Point, Cape May City, the Wildwood Island communities, Lower Township, and Middle Township (Lacombe and Carleton, 1992, 2002). Saltwater intrusion has forced the closure of at least 20 public- and industrial-supply wells and more than 100 domestic-supply wells since the 1940s. State and local water-resource planners recognize that saltwater intrusion toward the Cape May peninsula is likely to continue if withdrawals from the aquifer system are maintained at current rates or are increased to meet projected increases in water demands. In response to these concerns, alternative sources of water supply and new approaches for water-supply management at county and local levels are being evaluated and implemented.

One of the communities most affected by saltwater intrusion is Cape May City, where the history of saltwater contamination has been well documented (Gill, 1962; Lacombe and Carleton, 1992, 2002; Blair and others, 1999). The primary source of freshwater for the city has been ground water withdrawn from five wells that tap the Cohansey aquifer (fig. 26). Pumping from these wells has led to saltwater contamination in four of the wells (fig. 27). The city's first two wells were installed in 1940 and 1945 and were located within 3,400 ft of the Atlantic Ocean shoreline (Blair and others, 1999). However, by 1950, saltwater contamination had forced the first well to be removed from service, and a third well was installed further inland to replace the contaminated well. Chloride concentrations in water from well 2 increased steadily through the 1950s, and by 1963, the chloride concentration in the well was 200 mg/L. Two additional wells (4 and 5) were installed in 1965 landward of the first three wells as the quality of water in well 2 deteriorated and as demand for additional water increased. Chloride concentrations in water from well 3 increased slowly from about 20 to 50 mg/L from

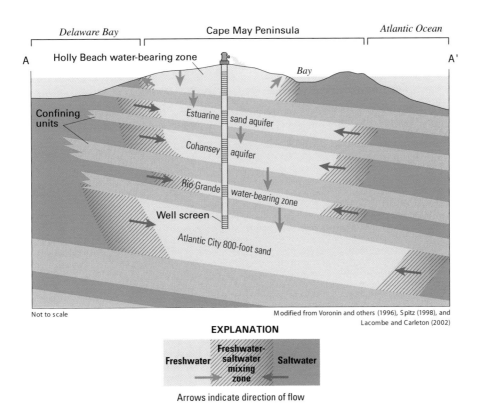

Figure 25. Generalized hydrogeologic section through the Cape May peninsula, New Jersey, showing the approximate extent of freshwater and saltwater and directions of ground-water flow in each aquifer. Ground-water withdrawals are shown schematically by a single well with screened intervals in each of the aquifers. Withdrawals from the wells have lowered ground-water levels and drawn the freshwater-saltwater interface landward. Line of section shown on figure 24.

1950 to the mid-1960s, but then increased to more than 100 mg/L in the 1970s after wells 1 and 2 had been removed from service and their withdrawals shifted to well 3. With the increased use of wells 4 and 5 since the 1970s, chloride concentrations have increased at well 4 from about 20 mg/L in 1965 to more than 100 mg/L beginning about 1985.

From 1985 to 1998, the city depended on well 5 for the majority of its year-round water supply. Withdrawals from well 3 were restricted to periods of peak water demand, and the water that was withdrawn from the well was blended with that from wells 4 and 5 to produce water of acceptable quality (Blair and others, 1999). In addition, since 1994 the city has operated a program of aquifer storage and recovery at well 4, in which water purchased from the Lower Township Municipal Utilities Authority is injected into the well during the off-season and withdrawn from the well during the summer tourist season.

During the mid-1990s, engineers working with the city had determined that projected population growth would result in a peak water demand that exceeded available supply by the year 2000 (Metcalf & Eddy, Inc., 1996). In response, the city sought an economical alternative to the existing water-supply system that would allow further expansion of tourism, would protect the Cohansey aquifer from saltwater intrusion, and could be implemented in a reasonable time period (Blair and others, 1999). The alternative that was judged most viable was one that involved the combined use of existing sources with a new source derived from desalinated brackish ground water. In this alternative, as much as 2 Mgal/d of brackish water would be pumped from two new supply wells installed in the Atlantic City 800-ft sand aquifer, which underlies the Cohansey aquifer and is brackish in the Cape May City area. The desalinated water would be used with freshwater obtained from continued operation of the aquifer storage and recovery system at well 4 and continued use of well 5 during periods of peak demand. The desalinated water is anticipated to reduce the city's reliance on wells 3 and 5 in the Cohansey aquifer and on water imported from Lower Cape May Township. Use of the brackish-water wells also will limit withdrawals from the Cohansey aquifer to periods of peak demand, which should reduce the rate of saltwater intrusion into the aquifer.

The desalination system began operation in 1998 after installation of wells 6 and 7 (fig. 26) and construction of a 2 Mgal/d-capacity desalination plant at the city's Canning House Lane Water Works facility. Brackish ground water is desalinated by use of reverse-osmosis membrane-filtration systems designed to treat brackish water containing up to 2,000 mg/L dissolved solids. Currently (2003), the systems treat about 0.5 Mgal/d on a typical day (David A. Carrick, City of Cape May Water Utility, New Jersey, oral commun., June 2003).

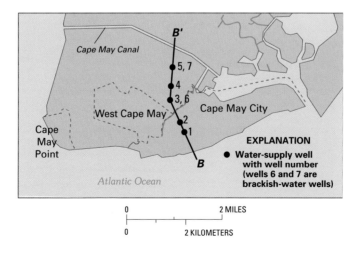

Figure 26. Cape May City water-supply wells and line of hydrogeologic section B-B′.

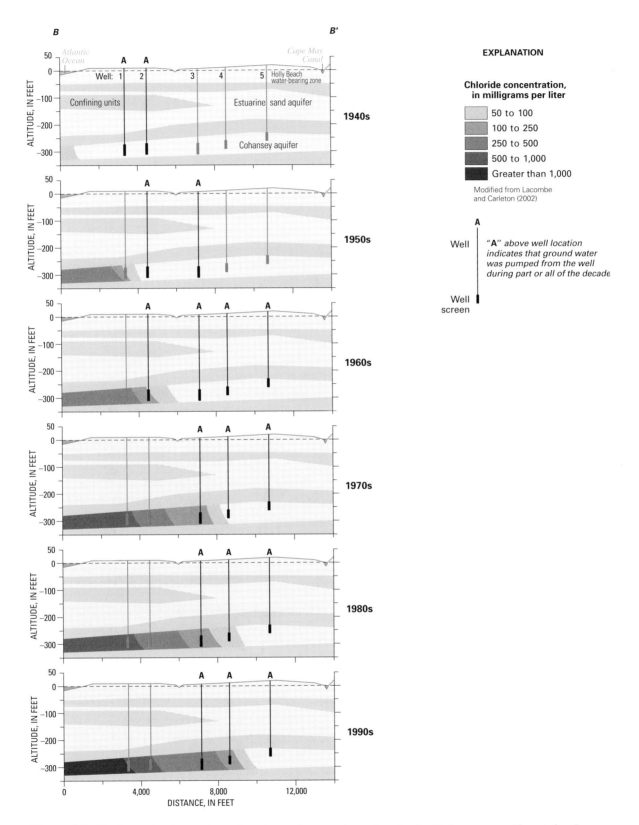

Figure 27. Hydrogeologic sections showing saltwater intrusion in the Cohansey aquifer at the Cape May City, New Jersey well field, 1940s to 1990s. Line of section shown on figure 26.

Saltwater Intrusion from the Delaware River During Drought—Implications for the Effects of Sea-Level Rise on Coastal Aquifers

Many shallow aquifer systems along the Atlantic coast are in direct hydraulic connection with streams and other surface waters that are often sources of recharge to the underlying ground-water system. This is the case in the Camden area of New Jersey, where the shallowest parts of the area's aquifer system are in connection with the Delaware River. Freshwater flowing seaward in the river mixes with saltwater from Delaware Bay in a transition zone that is characterized by an increase in the river's chloride concentration similar to the transition zone between freshwater and saltwater in coastal aquifers. The location in the river where the chloride concentration is 250 mg/L is called the "salt line" or "salt front" and is monitored by the Delaware River Basin Commission (Delaware River Basin Commission, 2001). The salt front's position is not stagnant, but moves upstream and downstream in response to tides, rainfall, and freshwater releases from reservoirs within the Delaware River Basin. During periods of average rainfall and streamflow, the position of the salt front is near Wilmington, Delaware (fig. 28A), about 30 mi downstream from Camden; however, during the summer and other periods of reduced flow in the river, the salt front migrates further upstream toward Philadelphia and Camden.

Ground-water withdrawals provide most of the potable water supply for the Camden area. Before these withdrawals began, ground water from the aquifers discharged to the river. However, because of the withdrawals, the direction of flow between the aquifer system and river has been reversed, and in many places river water now is drawn into the aquifer system. Under normal, or average, flow conditions, water in the river is fresh in the Camden area. However, during a period of severe drought that occurred in November and December 1964, freshwater flows to the river were extremely low, and the salt front migrated above Camden to its maximum recorded upstream position (fig. 28A). Ground-water withdrawals in the Camden area during the 1964 drought are thought to have drawn a slug of saltwater from the temporarily saline reaches of the river into the aquifer system (Navoy, 1991; Navoy and Carleton, 1995). Soon after the saltwater-contamination event occurred, higher-than-normal chloride concentrations were measured in ground water pumped from supply wells near the river (fig. 28B). Chloride concentrations at the wells generally peaked in 1965 or later, but because the sampling frequency was low, the peak concentration and timing of contamination at the wells could not be clearly established. The highest chloride concentration measured was about 80 mg/L, whereas the background chloride concentration of water in the aquifer system near the Delaware River is about 10 to 20 mg/L.

Although chloride concentrations in the Camden area supply wells remained below potability standards during and after the short drought-related saltwater-intrusion event of 1964, contamination of the wells demonstrates the vulnerability of coastal aquifers to saline surface water (Navoy, 1991; Navoy and Carleton, 1995). Sea-level rise could have a similar effect on coastal aquifers because as the sea level rises, saltwater would be pushed landward and further upstream into coastal estuaries. The potential consequences of sea-level rise on coastal aquifers is of concern because global (eustatic) mean sea level has increased at an average rate of 1 to 2 millimeters per year (mm/yr) during the 20th century (a total of about 4 to 8 inches), and this increase has been attributed to an increase in the Earth's temperature during the same period (Intergovernmental Panel on Climate Change, 2001). Global warming drives sea-level rise by thermal expansion of seawater and by melting of ice caps, ice fields, and mountain glaciers. Moreover, recently developed climate models suggest that global sea level may increase from 0.09 to 0.88 meter (m) (0.3 and 2.9 ft) from 1990 to 2100 (Intergovernmental Panel on Climate Change, 2001).

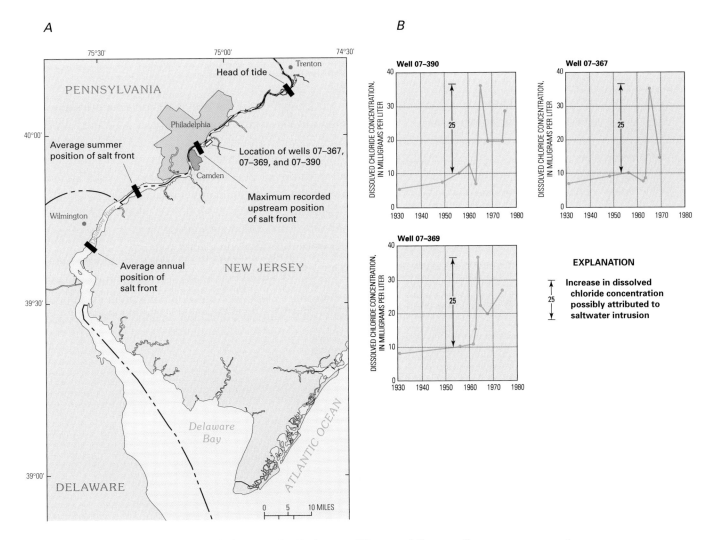

Figure 28. (A) Location of the salt front in the Delaware River and Estuary for average annual, average summer, and maximum recorded upstream positions; (B) dissolved-chloride concentrations of water from selected water-supply wells near the Delaware River in Camden, New Jersey, 1930 to 1980.

Vertical land movements in the earth's crust locally alter the rate of sea-level change from the global value, but usually by no more than a factor of 2 or 3; along the conterminous United States, local rates of sea-level rise average about 2 to 3 mm/yr, with rates of as much as 6 mm/yr or more in some specific areas (Douglas and others, 2001). The variability in the local rate of land movement is reflected in different rates of relative sea-level rise (that is, relative to a fixed datum on land) measured at individual gaging stations along the Atlantic coast. At the Boston, Massachusetts, tide gage, for example, relative sea level rose at a rate of about 2.5 mm/yr from 1921 to 2000 (fig. 29).

Sea-level rise could affect coastal aquifers in several ways (Navoy, 1991; Nuttle and Portnoy, 1992; Ayers and others, 1994; Navoy and Carleton, 1995; Oude Essink, 1999; Sherif, 1999; Sherif and Singh, 1999; and Douglas and others, 2001). Perhaps most fundamentally, a landward movement of seawater would push saltwater zones in coastal aquifers landward and upward, which could accelerate rates of saltwater intrusion into aquifers already experiencing saltwater contamination. Rising sea levels also might cause upstream migration of saltwater in coastal estuaries, inundation of low-lying areas including wetlands and marshes, and submergence of coastal aquifers. In some areas, sea-level rise would erode beaches and bluffs, leading to shoreline retreat, narrowing of aquifers, and diminished areas of aquifer recharge. Sea-level rise also might cause increases in coastal ground-water levels, because of the overall rise in the position of the freshwater-saltwater interface. For example, McCobb and Weiskel (2003) report an increase of 2.1 mm/yr in the average ground-water level measured from 1950 to 2000 at an observation well on the outer part of Cape Cod, Massachusetts (well TSW1 shown in fig. 11). Because this increase is similar to the relative sea-level rise measured at the Boston tide gage from 1921 to 2000 (2.5 mm/yr), McCobb and Weiskel (2003) hypothesize that the ground-water-level increase may reflect sea-level rise.

Although sea-level rise could increase saltwater intrusion into coastal surface and ground waters, landward saltwater movement also will depend in part on changes in precipitation, runoff, and recharge that may occur within coastal watersheds. For example, increased freshwater runoff could counterbalance the landward movement of saltwater. Moreover, should saltwater intrusion into coastal aquifers occur in response to sea-level rise, it is likely that some aquifers may require hundreds to thousands of years to re-equilibrate to changes in sea level, such as has occurred in parts of the Northern Atlantic Coastal Plain aquifer system where freshwater-saltwater interfaces appear to be still responding to sea-level increases that began at the end of the last ice age.

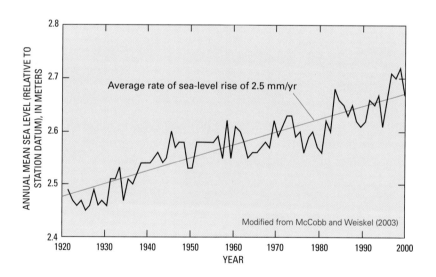

Figure 29. Annual mean sea-level rise at the Boston, Massachusetts, tide gage, 1921–2000. Mean relative sea level at the gaging station has increased during the period at an average rate of about 2.5 millimeters per year (mm/yr). This increase reflects both the global (eustatic) sea-level rise as well as local movements in the earth's crust. (Data from National Oceanic and Atmospheric Administration, 2001.)

Multifaceted Strategy for Managing Saltwater Intrusion in Coastal Georgia

Ground water withdrawn from the Upper Floridan aquifer is the principal source of water supply for 24 counties of coastal Georgia (fig. 30), an area of rapid population growth, increased tourism, and sustained industrial activity. The Upper Floridan aquifer is an extremely permeable, high-yielding aquifer that was first developed in the 1800s and has been used extensively in the area ever since. Total ground-water withdrawals in coastal Georgia were nearly 350 Mgal/d in 1997 (fig. 31); most of the water was used for industry, public supply, and irrigation. These withdrawals have resulted in substantial water-level declines in the coastal zone (fig. 30), decreased fresh ground-water discharge to springs and other surface-water features, and saltwater contamination at the northern end of Hilton Head Island, South Carolina, and in Brunswick, Georgia (figs. 32 and 33). This contamination is restricting further development of the Upper Floridan aquifer in coastal Georgia and adjacent parts of South Carolina and Florida and has created competing demands for the limited freshwater supply (Krause and Clarke, 2001). In response to the contamination and related water-supply issues, the Georgia Environmental Protection Division (GaEPD) has developed an interim strategy for managing saltwater intrusion in the Upper Floridan aquifer in the 24 coastal-area counties. The State's objective is to stop the intrusion of saltwater before municipal supply wells on Hilton Head Island and Savannah are contaminated and to prevent the existing saltwater contamination problem in Brunswick from worsening (Georgia Environmental Protection Division, 1997).

The multifaceted management strategy calls for a mix of regulatory and nonregulatory approaches for controlling saltwater intrusion. As part of the interim strategy, the GaEPD has restricted permitted withdrawals of water from the Upper Floridan aquifer in parts of the coastal area (including the Savannah and Brunswick areas) to 1997 rates, has restricted additional permitted pumpage in all 24 coastal counties to 36 Mgal/d above 1997 rates, and has encouraged and promoted conservation and reduced ground-water use wherever feasible throughout southeast Georgia. In addition, each of the coastal counties is required to develop comprehensive water-supply plans to manage the Upper Floridan aquifer, including identifying possible alternative sources of water to the Upper Floridan.

Flowing wells such as this one were common throughout the Georgia coastal area prior to large-scale ground-water development.

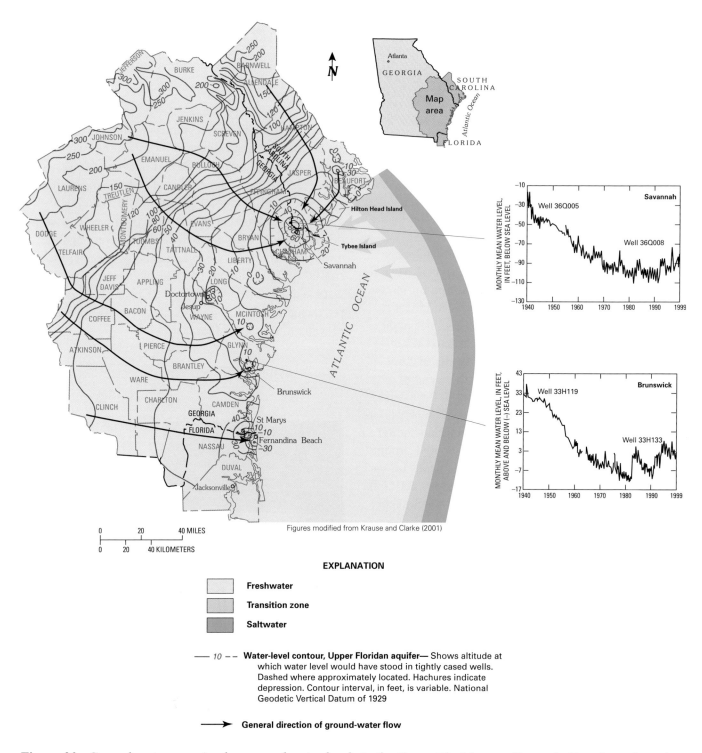

Figure 30. Ground-water pumping has caused water levels in the Upper Floridan aquifer to decline throughout the coastal area of southeast Georgia, and has created water-level depressions in areas of heavy, concentrated pumping, such as the Savannah, Brunswick, and St Marys-Fernandina Beach, Florida, areas. These water-level declines are illustrated by the two inset graphs of ground-water levels monitored at observation wells in the Savannah and Brunswick areas. Water-level declines in the Upper Floridan aquifer have caused saltwater contamination in at least two locations—Brunswick, Georgia, and Hilton Head Island, South Carolina (Krause and Clarke, 2001).

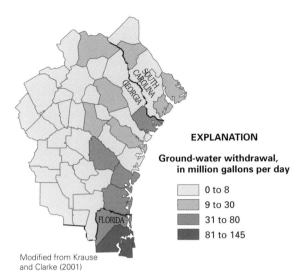

Figure 31. Ground-water withdrawals in the 24-county area of coastal Georgia served more than 500,000 people in 1997. Photograph shows an industrial water user, Savannah, Georgia.

An important element of the interim strategy has been a program of scientific and engineering studies to support development of GaEPD's final strategy to protect the Upper Floridan aquifer from saltwater contamination, which has been proposed for implementation in January 2006. The Coastal Georgia Sound Science Initiative is a collaborative effort among governmental, university, and private organizations to better understand the processes of ground-water flow and saltwater intrusion in the coastal zone of Georgia and adjacent parts of South Carolina and Florida, to monitor saltwater conditions in the coastal zone, and to evaluate alternative sources of water to the Upper Floridan aquifer. Some of the activities of the Sound Science Initiative are:

- Construction of temporary monitoring wells offshore from the Savannah-Hilton Head Island area to delineate areas where saltwater enters the Floridan aquifer system, such as submarine exposures or paleochannels;

- Development and expansion of a network of ground-water monitoring wells and an associated information database to measure and report changes in ground-water levels and chloride concentrations;

- Development of ground-water flow and solute-transport models to investigate the paths and rates of ground-water flow and saltwater intrusion in the Upper Floridan aquifer and to simulate alternative water-management strategies (Box E);

- Evaluation of alternative and supplemental sources of water such as seepage ponds, rivers and streams, reclaimed water, water desalinated by reverse osmosis, and ground water withdrawn from aquifers other than the Upper Floridan; and

- Feasibility studies and assessments of engineered and nonengineered methods that might be used to prevent saltwater intrusion.

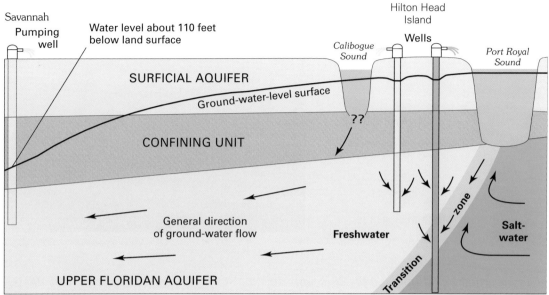

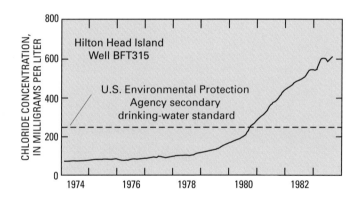

Figures modified from Krause and Clarke (2001)

Figure 32. Ground-water-level declines caused by pumping on Hilton Head Island, South Carolina, and in the Savannah, Georgia, area have resulted in saltwater contamination along the northern part of Hilton Head Island. The contamination probably is the result of lateral intrusion of seawater combined with some downward leakage of seawater to the Upper Floridan aquifer where the overlying confining unit is thin or absent (Krause and Clarke, 2001). Erosion of the confining unit in the vicinity of Port Royal Sound may have occurred during the last ice age when sea level was lower and streams cut down into the rocks of the Upper Floridan aquifer. Additional scouring of the confining unit may have resulted from large, present-day tidal fluctuations and perhaps also from dredging for phosphate ore in the late 1800s (Landmeyer and Belval, 1996). The graph shows chloride concentration of ground water in well BFT315 located at the northern end of Hilton Head Island.

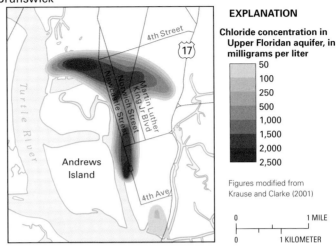

Downtown Brunswick

EXPLANATION

Chloride concentration in Upper Floridan aquifer, in milligrams per liter
- 50
- 100
- 250
- 500
- 1,000
- 1,500
- 2,000
- 2,500

Figures modified from Krause and Clarke (2001)

1962 Chloride contamination covers a 0.4-square mile area. Measured chloride concentrations are as high as 860 milligrams per liter

1991 Chloride contamination covers a 2.9-square mile area. Measured chloride concentrations exceed 2,500 milligrams per liter at two locations

Figure 33. Saltwater contamination at Brunswick, Georgia, is the result of upward intrusion of saline water from the deepest zone of the Lower Floridan aquifer (called the Fernandina permeable zone) into the freshwater parts of the Lower and Upper Floridan aquifers (Krause and Clarke, 2001).

Drilling offshore of Tybee Island, Georgia, is helping to delineate the location of the freshwater-saltwater interface in the Upper Floridan aquifer as well as areas where saltwater is entering the Floridan aquifer system. The offshore data collection from a U.S. Army Corps of Engineers drill rig is part of the Coastal Georgia Sound Science Initiative.

Numerical Modeling of Coastal Aquifers

Throughout this report, references are made to the development of numerical models that simulate coastal aquifers. Numerical modeling has emerged over the past 40 years as one of the primary tools that hydrologists use to understand ground-water flow and saltwater movement in coastal aquifers. Numerical models are mathematical representations (or approximations) of ground-water systems in which the important physical processes that occur in the systems are represented by mathematical equations.[1] These equations are solved by mathematical techniques (such as finite-difference or finite-element methods) that are implemented in computer codes. The primary benefit of numerical modeling is that it provides a means to represent, in a simplified way, the key features of what are often complex systems in a form that allows for analysis of past, present, and future ground-water flow and saltwater movement in coastal aquifers. Such analysis is often impractical, or impossible, to do by field studies alone.

Numerical models have been developed to simulate ground-water flow solely or ground-water flow in combination with solute transport (the movement of chemical species through an aquifer). For a number of reasons, numerical models that simulate ground-water flow and solute transport are more difficult to develop and to solve than those that simulate ground-water flow alone. Coastal aquifers are particularly difficult to simulate numerically because the density of the water and the concentrations of chemical species dissolved in the water can vary substantially throughout the modeled area. To address these difficulties, one of two approaches generally is used to simulate freshwater-saltwater interactions (Reilly and Goodman, 1985; Reilly, 1993). In the first approach, the freshwater and saltwater zones are assumed to be immiscible (that is, they do not mix) and separated by a sharp interface. In the second approach, the freshwater and saltwater are considered to be a single fluid having a spatially variable salt concentration that influences the fluid's density; this approach is referred to as density-dependent ground-water flow and solute-transport modeling.

A recent example of the use of a numerical model to evaluate a coastal ground-water system is one developed to quantify the rates and patterns of submarine ground-water discharge to Biscayne Bay in southeastern Florida (fig. E–1). The bay is a coastal barrier-island lagoon that relies on substantial quantities of freshwater to sustain its estuarine ecosystem. The two most important sources of freshwater to the bay are thought to be drainage canals and ground-water discharge from the Biscayne aquifer (Langevin, 2001). Although canal discharges are routinely measured, few studies have attempted to quantify the contribution from ground water. To better understand the interaction between the Biscayne aquifer and Biscayne Bay, a numerical model was developed to simulate conditions during 1989 to 1998.

The numerical model was developed on the basis of field investigations that provided information on the structure and hydraulic properties of the aquifer; interactions of the ground-water system with the bay and drainage canals; and hydraulic stresses to the aquifer, which include recharge from precipitation, evapotranspiration, and ground-water withdrawals. The computer code that was used in the study, which is called SEAWAT (Guo and Bennet, 1998; Guo and Langevin, 2002), is based on the second approach for simulating coastal freshwater-saltwater interactions (density-dependent ground-water flow and solute transport). One of the outputs of the model was a calculated distribution of ground-water salinity within the modeled area (fig. E–1). In general, the model reasonably simulated the inland extent of saltwater within the aquifer, particularly south of the Miami Canal. The average rate of fresh ground-water discharge to the bay calculated by the model for 1989 to 1998 was about 53 Mgal/d. Overall, ground-water discharge to the bay was only about 6 percent of the measured canal discharge to the bay for the same 10 years.

---

[1] Several textbooks and articles are available on numerical modeling of ground-water flow systems, including those by Anderson and Woessner (1992) and Konikow and Reilly (1999), which were used as references for this discussion. A review of several computer codes that have been developed for saltwater-intrusion modeling is provided by Sorek and Pinder (1999).

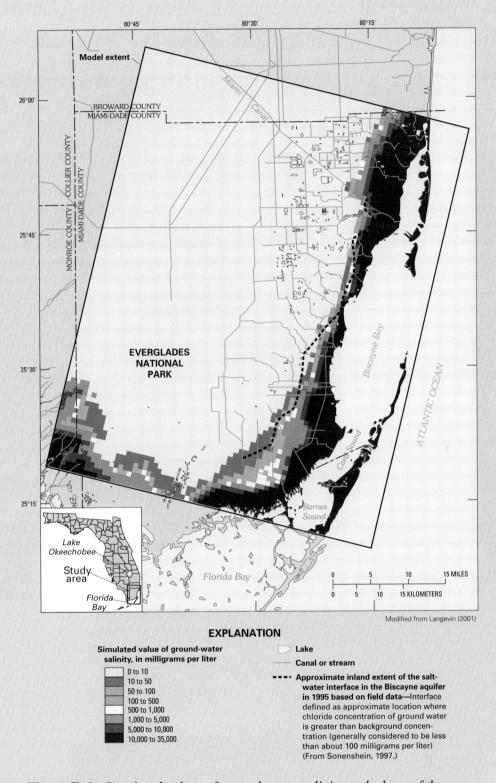

Figure E–1. Simulated values of ground-water salinity at the base of the Biscayne aquifer.

Vertical Migration of Saline Water Along Preferential Flow Conduits in the Floridan Aquifer System—Implications for Saltwater Monitoring Networks

The Floridan aquifer system is the major source of water supply in southeastern Georgia and northeastern Florida. Ground-water withdrawals from the aquifer system in the three northeastern-most coastal counties of Florida alone were 217 Mgal/d in 1995 (Marella, 1999). These withdrawals have caused long-term water-level declines in several areas of the region; in northeastern Florida, for example, water levels in the Floridan aquifer system have declined during the past 50 to 60 years at an average rate of about one-third to one-half foot per year (Spechler, 1994, 2001). Ground-water development in the region has resulted in a complex process of saltwater contamination by upward migration of saline water from the deepest zone of the aquifer system (called the Fernandina permeable zone) to overlying freshwater zones along structural anomalies in the geologic framework of the aquifer system that provide preferential conduits for ground-water flow (figs. 34 and 35). This type of contamination has occurred at Brunswick, Georgia (fig. 33), at Fernandina Beach, Florida, and near Jacksonville in Duval County, Florida (fig. 36).

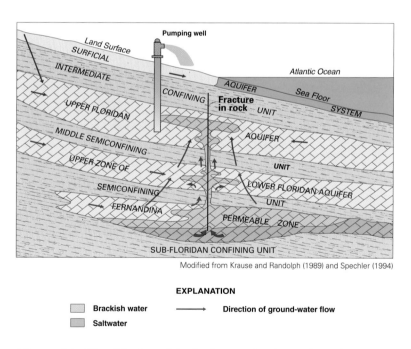

Modified from Krause and Randolph (1989) and Spechler (1994)

Figure 34. Simplified model of saltwater leakage along fractures in the Floridan aquifer system in southeastern Georgia and northeastern Florida. The fractures and other structural anomalies such as faults and dissolution-collapse features provide preferential conduits for saline water in the Fernandina permeable zone to flow upward into freshwater zones of the overlying aquifers in response to ground-water pumping in the upper aquifers.

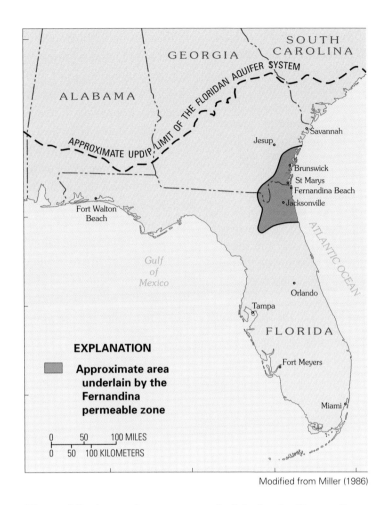

Figure 35. Approximate area underlain by the Fernandina permeable zone, southeastern Georgia and northeastern Florida.

The Fernandina permeable zone is an extremely permeable, partly cavernous water-bearing zone in the Lower Floridan aquifer. Chloride concentrations of water pumped from the zone have been reported to be as high as 16,800 mg/L in St. Johns County, Florida, and 33,000 mg/L in the Brunswick area of Georgia (Spechler, 1994; Krause and Clarke, 2001). The Fernandina is overlain by a low-permeability confining unit that, in most places, effectively separates it from shallower permeable strata (Miller, 1986). However, in a number of locations, isolated geologic structures have breached the Floridan aquifer system and created conduits that allow saltwater to migrate upward from the Fernandina permeable zone to overlying freshwater aquifers, especially in areas where ground-water levels have been lowered by pumping (fig. 34). Geologic evidence indicates that the conduits are most likely fractures, joints, or faults in the rocks that have been enlarged by dissolution of the carbonate strata in the fractured zone (Leve, 1983; Krause and Randolph, 1989; Maslia and Prowell, 1990; Spechler, 1994, 2001; Phelps and Spechler, 1997; Phelps, 2001). In some locations, such as northeastern Florida, dissolution of the rocks may have resulted in the subsequent collapse of overlying strata, further enhancing the hydraulic connection between freshwater and underlying saltwater (Spechler, 1994, 2001). Once the saltwater reaches the freshwater zones, it moves laterally downgradient within the freshwater zones toward areas of pumping (fig. 34).

The upward migration of saline water along isolated conduits of preferential flow has created localized, anomalous patterns of saltwater contamination that often are difficult to discern. For example, in Brunswick, Georgia, the highest chloride concentrations have been found at the locations of the points of intrusion of the contaminated water into the freshwater aquifer—more than 10,000 ft south of the area of maximum ground-water withdrawals (Maslia and Prowell, 1990). The preferential flow paths also make it difficult to design monitoring systems to provide early warning of saltwater contamination or to develop new water-supply wells, because the conduits that provide the pathways for saltwater movement are difficult to map from the land surface. Conventional saltwater monitoring systems are not as useful in this hydrogeologic setting because such systems are designed on the assumption that the location of the saltwater contaminant and the most likely pathways it can take are known with some degree of confidence. Monitoring for saltwater contamination in this setting requires continued sampling of existing water-supply and monitoring wells in the freshwater aquifers, including intensive monitoring of selected wells in which chloride concentration increases have been detected, to provide indications of trends of deteriorating water quality (Phelps and Spechler, 1997). Chloride concentrations of ground water in the Brunswick area, for example, have been sampled since the 1960s, providing a long-term record of changes in the water-quality conditions of the Floridan aquifer system in that area. Moreover, the use of seismic and borehole geophysical methods may increase knowledge of the locations of the buried conduits that facilitate saltwater movement; previous seismic geophysical investigations have revealed a number of structural anomalies in northeastern Florida (Spechler, 1994; Odum and others, 1999; Kindinger and others, 2000).

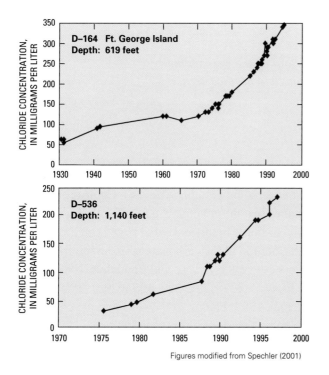
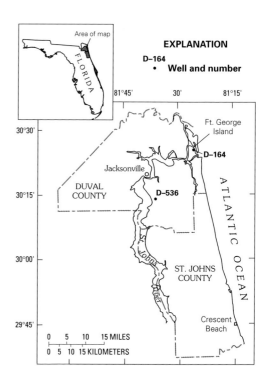

Figure 36. Graphs show steady increases in chloride concentrations in water from two wells tapping the Floridan aquifer system in northeastern Florida. Elevated chloride concentrations have been observed in more than 70 wells in the area that tap the Upper Floridan and the upper zone of the Lower Floridan aquifers, in some instances as much as 14 miles inland of the coast (Spechler, 2001).

Vertical Migration of Saltwater Across Interconnected Aquifers in Water-Supply Wells—Contamination and Abatement, West-Central Florida

Open boreholes and improperly constructed or corroded wells can be pathways for interaquifer saltwater intrusion when flow is induced by hydraulic-head gradients within the boreholes or wells, and when there is a source of saltwater that can be transported from one aquifer to another (fig. 37). Several studies in Florida have reported incidences of the movement of saltwater and other poor-quality waters within the boreholes of abandoned irrigation wells, which has resulted in deterioration of water quality in vertically adjacent aquifers. For example, a recent study in west-central Florida (fig. 38) determined the extent of interaquifer flow and the potential for ground-water quality deterioration as a result of flow in wells open to multiple aquifers containing water of varying quality. Because of its high productivity, the Upper Floridan aquifer is the most widely used of three interconnected aquifers in the area—the surficial aquifer, intermediate aquifer system, and Upper Floridan aquifer (fig. 37) (Metz and Brendle, 1996). The three aquifers are separated from each other by intervening confining units that, because of their relatively low permeability, limit the amount of upward and downward leakage among the aquifers. The significance of the Upper Floridan aquifer as a source of potable water diminishes as the water quality degrades in the southern and western parts of the study area. In these areas, the importance of the intermediate aquifer system as a source of water increases. Within the study area, poor-quality water refers to water having concentrations of chloride, sulfate, and total dissolved solids that exceed 250, 250, and 1,000 mg/L, respectively.

Thousands of deep water wells were drilled into the Upper Floridan aquifer in west-central Florida from 1900 to the early 1970s for irrigation. Most of the early irrigation wells also were open to the intermediate aquifer system. Usually, the wells were completed with a short length of steel casing through the surficial aquifer and then were open to the two lower aquifers (fig. 37). These open wells, which can be many hundreds of feet in length, provide direct conduits for water to flow upward or downward across the confining units, thus short-cutting the slower route of leakage through the confining units (Metz and Brendle, 1996). Approximately 8,000 wells were reported or estimated to be open to the intermediate aquifer system and the Upper Floridan aquifer in the study area.

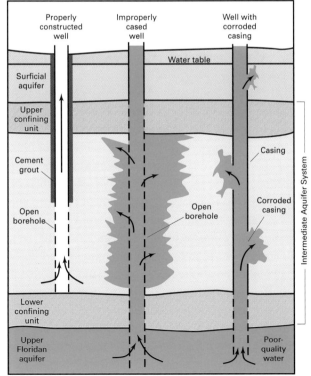

Figure 37. Contamination by saltwater and other poor-quality waters can occur through failed, uncased, or improperly constructed wells that create a conduit for flow between aquifers of differing water quality. A properly constructed well open to a single aquifer is shown on the left. Contamination of the intermediate aquifer system occurs by upward movement of poor-quality water from the Upper Floridan aquifer within the improperly cased and corroded wells.

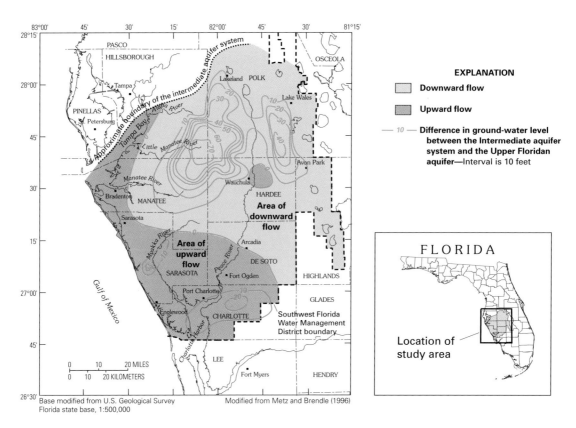

Figure 38. Difference in ground-water levels between the intermediate aquifer system and underlying Upper Floridan aquifer, west-central Florida, September 1993. Differences in ground-water levels were measured at the end of the summer rainy season at a time when the aquifers generally are unstressed by irrigation pumpage.

Areas of upward and downward flow between the intermediate aquifer system and the Upper Floridan aquifer were identified on the basis of ground-water-level measurements made throughout the multi-county area (fig. 38). Ground-water levels in the intermediate aquifer system were as much as 70 ft higher than those in the underlying Upper Floridan aquifer in areas inland from the coast; however, near the coast and in the southern part of the study area, ground-water levels were as much as 20 ft higher in the Upper Floridan aquifer than in the overlying intermediate aquifer system (fig. 38). It was estimated that a total of 85 Mgal/d flowed from the Upper Floridan aquifer to the overlying fresher water zones in the intermediate aquifer system through interaquifer wells. In the majority of the area of upward flow, concentrations of chloride, sulfate, and dissolved solids in the Upper Floridan aquifer exceed recommended or permitted drinking-water standards, and the upward flow is contaminating the intermediate aquifer system. In the area of downward flow, an estimated 127 Mgal/d of potable water was moving through interaquifer wells from the intermediate aquifer system and artificially recharging the underlying Upper Floridan aquifer.

In 1974, the Southwest Florida Water Management District began the Quality of Water Improvement Program to restore hydrologic conditions altered by improperly constructed wells through a process of plugging of abandoned artesian wells (Southwest Florida Water Management District, 2002). One of the goals of the program is to re-establish confinement between aquifers by plugging the sections of well bores responsible for hydraulic connection between one or more aquifer zones. Confinement is re-established by plugging the borehole with cement or bentonite (clay) sealing materials to halt interaquifer contamination. The program also is concerned with identifying uncontrolled (free-flowing) wells that discharge poor-quality water at land surface. As of October 2001, more than 5,200 wells had been inspected and nearly 3,000 plugged since the program began (Southwest Florida Water Management District, 2002).

Saltwater Intrusion in Southeastern Florida—Response of the Biscayne Aquifer to Large-Scale Drainage and Ground-Water Withdrawals

Saltwater intrusion caused by large-scale drainage and ground-water withdrawals has been an issue of concern in southeastern Florida since the 1930s. The area is part of the region that once contained one of the largest wetlands in the continental United States, the Everglades, which extended across the southern part of the Florida Peninsula south of Lake Okeechobee (fig. 39). The prevalence of wetlands is the result of the region's abundant rainfall and low, flat terrain. Drainage of south Florida's watersheds began in the early 1880s with the construction of drainage canals to reclaim land north of Lake Okeechobee (McPherson and Halley, 1996). Beginning with the Miami River in 1903 and continuing into the 1980s, an extensive network of canals was constructed between the lake and the Atlantic Ocean to drain the Everglades for agricultural and urban development and, with time, to convey water across the region to meet agricultural and ecological needs, to control flooding, and to mitigate saltwater intrusion. Concurrently, a system of levees was constructed throughout the region, first to control flooding from Lake Okeechobee and later to impound water in large water conservation areas southeast of the lake for release during the dry seasons. The canal and levee systems enabled agriculture to develop on hundreds of thousands of acres south of Lake Okeechobee and contributed to the explosive population growth of the area from less than 4,000 inhabitants in 1900 to more than 3.8 million in 2000. Today, the canals and levees are part of a complex water-management system operated by the South Florida Water Management District to sustain the Everglades, meet agricultural and urban water demands, and control flooding and saltwater intrusion.

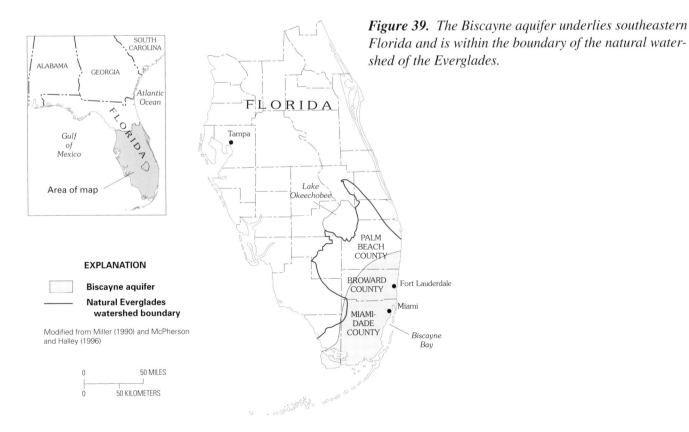

Figure 39. The Biscayne aquifer underlies southeastern Florida and is within the boundary of the natural watershed of the Everglades.

An important component of southeastern Florida's hydrologic system is the highly productive Biscayne aquifer (fig. 39), which is the source of water supply for more than 3 million people in Miami-Dade, Broward, and Palm Beach Counties. Total withdrawals from the aquifer were about 870 Mgal/d in 1995 (Marella, 1999); withdrawals from the aquifer in Miami-Dade County alone (550 Mgal/d) were greater than any other county along the Atlantic coast. The Biscayne aquifer underlies an area of about 4,000 mi^2 and extends beneath Biscayne Bay, from which its name was derived, and the Atlantic Ocean. The aquifer consists of highly permeable limestone and less-permeable sandstone and sand (Miller, 1990). The aquifer is underlain by about 600 to 1,000 ft of low-permeability, largely clayey deposits that hydraulically separate the Biscayne aquifer from the underlying Floridan aquifer system, which contains saltwater in southeastern Florida.

The general movement of water within the Biscayne aquifer is from inland regions to coastal discharge areas. Water in the aquifer is under unconfined conditions, and the water table responds quickly to recharge, evapotranspiration, and pumping from supply wells. Moreover, in most places the aquifer and canals are in direct hydraulic connection, which facilitates rapid interchange of water between the canals and ground-water system (fig. 40). This connection allows the canals to drain the aquifer when ground-water levels are greater than canal stages and to recharge the aquifer when ground-water levels are lower than canal stages. Prior to 1945, flow within the canals was uncontrolled, causing an overdrainage of the aquifer and periodically allowing seawater to move inland along the canals and subsequently into the ground-water system (fig. 41). Since 1946, gated structures have been constructed on the primary canals to prevent inland migration of seawater along the canals and to control canal stages and ground-water levels. During the wet season, the gates are opened to lower water levels and discharge excess surface and ground water to the ocean to prevent flooding. During the dry season, the gates are closed to raise canal stages above ground-water levels near the coast, inducing water to seep from the canals into the aquifer and retarding saltwater intrusion. In addition to providing drainage and preventing flooding, regulation of the canals provides a means to transfer water from inland areas to coastal reaches. In these areas, canal water is artificially recharged to the aquifer to raise coastal ground-water levels, control saltwater intrusion, and provide a source of water to coastal well fields.

Photograph courtesy of the South Florida Water Management District

Dam-like control structures such as this one on Snake Creek Canal have been constructed in near-coastal reaches of the major drainage canals. The structures are closed to prevent inland movement of saltwater up the canals during periods of lower precipitation.

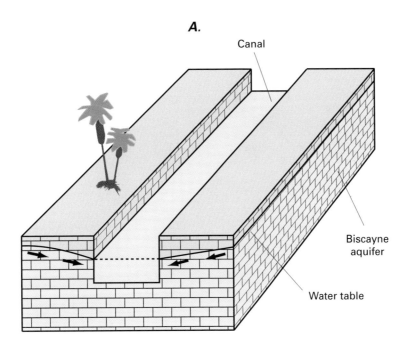

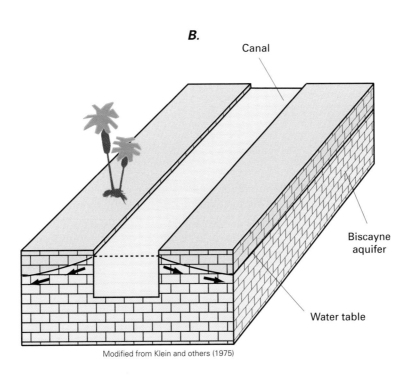

Figure 40. *In most places, there is a direct hydraulic connection between the Biscayne aquifer and canals that have been dug into it. This connection allows water to pass freely between the canals and the ground-water system. When the water level in the aquifer is higher than that in the canal, ground water moves toward the canal (A); when the water level in the canal is higher than that in the aquifer, water in the canal moves into the aquifer (B).*

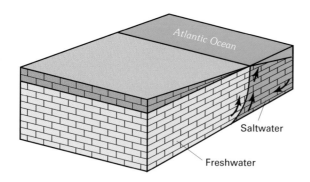

The freshwater-saltwater interface was nearly stable before coastal canals were built.

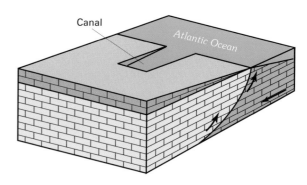

Uncontrolled tidal canals caused saltwater intrusion by lowering freshwater levels and providing open channels to the sea.

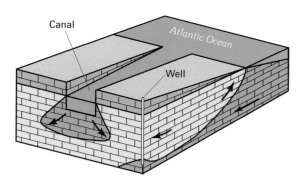

An uncontrolled canal that extended into an area of heavy pumping could convey saltwater inland to contaminate freshwater supplies.

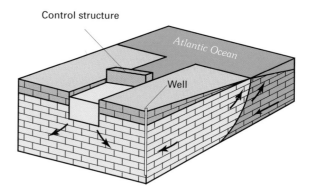

In contrast, a controlled canal provides a perennial supply of freshwater from upgradient areas to prevent saltwater intrusion and to recharge a well field.

Figures modified from Klein and others (1975)

Figure 41. Saltwater intrusion in southeastern Florida has been caused by the construction of drainage canals in addition to ground-water withdrawals for water supply. Initially, the canals were uncontrolled, which resulted in overdrainage of the aquifer and periodic movement of seawater inland along the canals and subsequently into the ground-water system. Since 1946, control structures have been placed on the canals to prevent inland migration of seawater, as well as to provide flood protection and artificial recharge to the aquifer.

Construction of the extensive drainage system and large-scale ground-water withdrawals have resulted in saltwater intrusion into coastal areas of the Biscayne aquifer. Canal drainage appears to have had the most widespread impact on saltwater intrusion (Sonenshein, 1997; R.A. Renken, U.S. Geological Survey, written commun., 2002). The canals have lowered ground-water levels and induced lateral saltwater intrusion along the coast, and, at least prior to the construction of the canal control structures, conveyed seawater inland to the ground-water system. Saltwater contamination began soon after the initiation of coastal drainage, forcing the closure of the Spring Gardens well field located about 1.5 mi from Biscayne Bay by 1925 and of the

Coconut Grove well field, about 1 mi from the bay, by 1941 (Parker and others, 1955). In addition, by 1946, thousands of private supply wells reportedly had been abandoned along the coastline in Miami-Dade County because of saltwater intrusion into the aquifer (Parker and others, 1955). In an effort to develop ground-water supplies that are unaffected by saltwater intrusion, there has been a general inland shift over time in the construction of new supply wells.

The landward limit of saltwater in the Biscayne aquifer in 1995 indicates that saltwater intrusion has occurred along much of the shoreline of the three-county area (fig. 42). Movement of the interface near Miami has been particularly well documented through the years (fig. 43). In 1904, prior to construction of the regional drainage network and water-supply wells, very high water levels in the Everglades and Biscayne aquifer kept the saltwater interface limited to a narrow band along the coastline and to short tidal reaches of coastal streams (fig. 43*A*). Beginning in 1909 with the extension of the Miami River by the Miami canal and continuing into the early 1940s, construction of drainage canals that lacked control structures and pumpage from coastal well fields resulted in lowered water levels in the Biscayne aquifer and substantial inland movement of saltwater along the coastline and into coastal tidal canals (fig. 43*B* and *C*). Drought conditions during the 1930s and 1940s, and particularly during 1943 to 1945, further exacerbated the saltwater-intrusion problem. Parker and others (1955) concluded that the major spread of saltwater contamination probably occurred from 1943 to 1946. They estimated that the rate of intrusion in the Silver Bluff area of Miami until 1943 had been approximately 235 ft/yr, but during a 27-month period that overlapped 1943–44, the saltwater front advanced at a rate of 890 ft/yr. Uncontrolled drainage in southeastern Florida was halted by the installation of canal control structures beginning in 1946. The inland migration of saltwater has slowed and in some areas reversed because of the effects of the control structures, as illustrated by the 1953, 1977, and 1995 positions of the saltwater front in the Miami area (fig. 43*D–F*).

Alteration of the south Florida hydrologic system over the past 100 years has had severe environmental effects beyond those of lowered ground-water levels and saltwater intrusion. These effects include the elimination of about one-half of the original extent of the Everglades and fragmentation of the remaining landscape, changes in the timing and duration of wetland flooding, degradation of water quality, and significant declines in populations of native plant and animal species (McPherson and Halley, 1996; U.S. Army Corps of Engineers and South Florida Water Management District, 1999). Recently, a plan to restore, preserve, and protect south Florida's ecosystems and to provide for other water-related needs of the region has been developed by the U.S. Army Corps of Engineers and the South Florida Water Management District working cooperatively with several other organizations. The Comprehensive Everglades Restoration Plan, which was approved by the U.S. Congress in December 2000, is a multicomponent project that would significantly modify existing water-management operations in the region. Although the primary focus of the plan is to recover the key ecological features of the Everglades and other south Florida ecosystems, the plan also has been designed to ensure continued water supply for urban and agricultural areas, to provide flood protection, and to prevent saltwater intrusion. One of the key components of the plan is the use of large-scale aquifer storage and recovery systems to store excess water during wet periods for recovery during dry periods (Box F).

Although saltwater intrusion is a continuing threat to the ground-water resources of southeastern Florida, the complex water-management system that has been developed across the region to maintain high water levels near the coast generally has controlled the intrusion. Moreover, monitoring networks have been established to track changes in ground-water levels and water-quality conditions in coastal areas. Future changes in the region's water-management operations that have been proposed in the Everglades restoration plan are likely to have consequences for the coastal ground-water system, including movement of the freshwater-saltwater interface. Continued monitoring and scientific analysis of the entire hydrologic system, as called for in the plan, will provide feedback on how these water-management changes affect the coastal ground-water system.

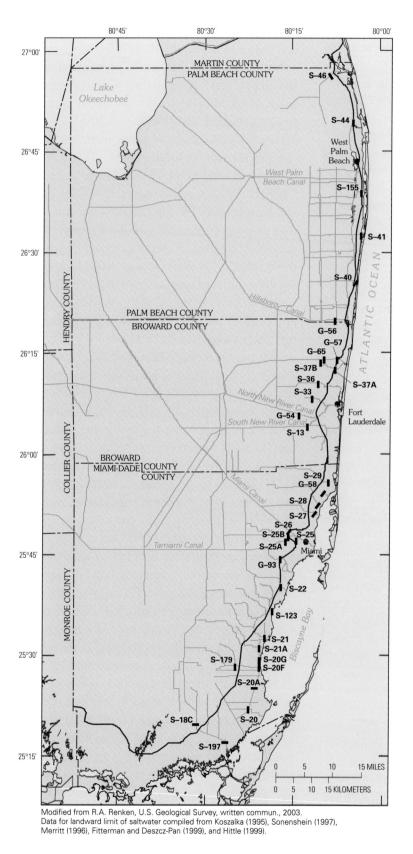

Figure 42. Landward limit of saltwater in the Biscayne aquifer, 1995.

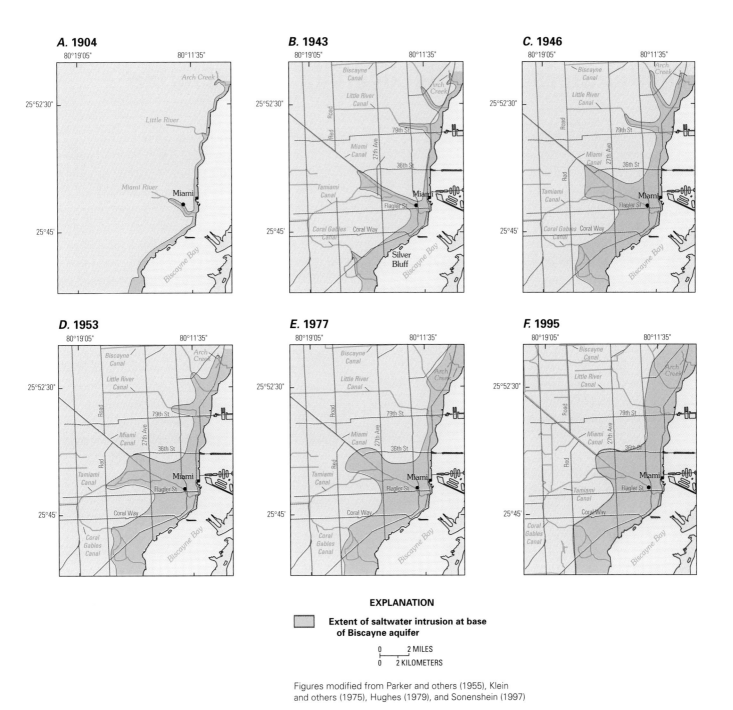

Figure 43. Saltwater intrusion in the Biscayne aquifer in Miami-Dade County near the Miami Canal, (A) 1904, (B) 1943, (C) 1946, (D) 1953, (E) 1977, and (F) 1995.

Aquifer Storage and Recovery in South Florida

Large-scale aquifer storage and recovery (ASR) systems have been proposed as part of the Comprehensive Everglades Restoration Plan (CERP) to store excess water during wet periods for recovery for ecological use and other water-resource needs during dry periods. Aquifer storage and recovery is a process by which water is recharged through wells to a suitable aquifer, stored for a period of time, and then extracted from the same wells when needed (Pyne, 1995). ASR systems have been proposed as part of the Everglades plan because, in comparison to surface-water storage, underground storage would limit evaporation of the stored water and require less land for storage (U.S. Army Corps of Engineers and South Florida Water Management District, 1999). Although ASR systems have been constructed at 26 sites in south Florida as of April 2001 (Reese, 2002), the scale of the ASR systems that are proposed for south Florida as part of the CERP—as much as 1.6 Bgal/d using about 330 wells—is unprecedented. The existing ASR systems consist predominantly of single-well systems located at municipal water-treatment plants, and the capacity of each well is substantially smaller than envisioned in the CERP.

The proposed storage zone for the ASR wells is the Upper Floridan aquifer, which is also the aquifer used for most of the existing ASR sites in south Florida (Reese, 2002). The Upper Floridan is continuous throughout south Florida, and its overlying confinement generally is good. Most of the existing ASR systems are located along the east and west coasts of south Florida, although those proposed as part of the Everglades plan would be constructed in inland areas, particularly around Lake Okeechobee.

One of the challenging aspects of ASR systems in south Florida is that the quality of the ambient water in the Upper Floridan aquifer is brackish to saline. Theoretically, recharge of freshwater into the saline-water aquifer creates a radial zone of mixing (the transition zone) around the well that separates the native saline water from the injected freshwater (fig. F–1). In reality, the shapes of the injected freshwater zone and of the mixing zone are highly dependent on the geology of the injection zone. Mixing between the injected freshwater and native saline water can reduce the amount of freshwater that is recovered during the withdrawal phase, and can lower the recovery efficiency of the ASR operation. Typically, the salinity of the injected and recovered water is closely monitored, usually on a daily basis. During recovery, salinity of the water increases with time, and recovery is terminated when the salinity of the water reaches a predetermined level (generally 250 mg/L or slightly higher if the recovered water is mixed with potable water at a water-treatment plant) (Reese, 2002). High salinity levels also can cause buoyant stratification, with the injected water rising over the native water. Other factors that affect the recovery efficiency of ASR systems include aquifer thickness and permeability, ambient hydraulic gradients, volume of injected water, duration and frequency of injection and withdrawal phases, duration of storage phases, the total number of completed injection-storage-withdrawal cycles, and well-bore plugging (Merritt, 1985; Pyne, 1995; Yobbi, 1997; Reese, 2002).

The large scale of the ASR systems that have been proposed as part of the CERP have raised a number of technical issues regarding full-scale implementation of the systems. These issues include: (1) the relative scarcity of information on the hydrogeology of the Upper Floridan aquifer in the areas of the proposed ASR wells; (2) the impacts of ground-water level (hydraulic head) buildup and decline to the regional ground-water flow system and on existing wells; (3) the possibility of rock fracturing in the aquifer and overlying confining unit caused by ground-water-head buildup; and (4) water-quality concerns regarding the suitability of the source waters for recharge and whether the quality of the recovered water would pose environmental or health concerns (National Research Council, 2001; Fies and others, 2002). In response to these concerns, a number of site-specific and regional-scale research activities, including ASR pilot projects, are planned or underway to better understand the hydrogeology and biogeochemistry of the natural system and the effects of ASR operation on the system, and to identify optimum well design and well-cluster configurations (U.S. Army Corps of Engineers and South Florida Water Management District, 1999; Fies and others, 2002).

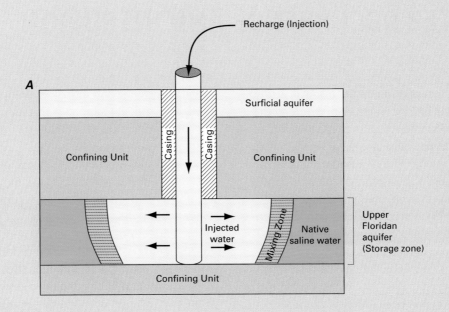
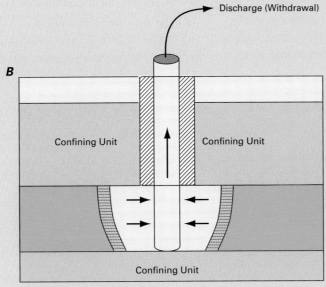

Figure F–1. Recharge (A) and discharge (B) phases for an idealized aquifer storage and recovery well in south Florida. A zone of injected water develops near the well during injection, and is separated from the native saline water by a mixing zone.

CHAPTER 3. DETECTING AND MONITORING SALTWATER OCCURRENCE AND INTRUSION

Since the early 1900s, numerous field studies have yielded a wealth of information on the occurrence and intrusion of saltwater in freshwater aquifers along the Atlantic coast. Field studies document the location and movement of saline water in coastal aquifers and, in a broader sense, are the basis for understanding the mechanisms that cause saltwater intrusion in different hydrogeologic settings (Reilly and Goodman, 1985; Bear and others, 1999). Several geochemical and geophysical techniques are used to directly or indirectly monitor saltwater in coastal aquifers. Because of the very high concentration of chloride in seawater (typically about 19,000 mg/L), the chloride concentration of ground-water samples has been the most commonly used indicator of saltwater occurrence and intrusion in coastal aquifers. However, other indicators of ground-water salinity, such as the total dissolved-solids concentration or specific conductance of ground-water samples, also are used frequently (Box G).

Several examples are provided in the next three sections that describe geochemical and geophysical techniques that are being used to detect and monitor saltwater occurrence and intrusion along the Atlantic coast. The geochemical techniques that are described include the commonly used approaches for characterizing saline water and less-frequently applied methods using isotope geochemistry. Geochemical isotopes are important tools in coastal-aquifer studies because they provide a means to differentiate among alternative sources of saline water. Although the case studies are focused primarily on methods for detecting and monitoring saltwater in coastal aquifers, it should be emphasized that a thorough understanding of the factors that affect the distribution and movement of saline water in a coastal aquifer also requires definition of the hydrogeologic framework, hydraulic properties, and physical boundaries of the aquifer and the distribution of ground-water levels and locations of ground-water withdrawals in the aquifer (Reilly, 1993). Recent summaries of many of the geochemical and geophysical approaches used in the study of freshwater-saltwater environments are provided by Jones and others (1999) and Stewart (1999), respectively.

Saltwater-Monitoring Networks Along the Atlantic Coast

Networks of saltwater-monitoring wells have been established in several States along the Atlantic coast to document the location and movement of saltwater in freshwater aquifers; selected networks are listed in table 1. These wells serve as early warning systems of the movement of saltwater into freshwater areas and provide data that can be used to measure the rate of saltwater movement. Water samples are collected for analysis of their chloride content or, in some cases, the total dissolved-solids content or specific conductance of the water. Frequently, water levels also are measured at the monitoring wells. The variability among the networks with respect to the number of wells in each network and their frequency of sampling is a function of several factors, which include the extent of active or potential saltwater intrusion in the aquifer, the areal extent of the aquifer and coastline that is being monitored, the distance of each well from the known or estimated saltwater front, and the availability of funding to support the network.

It should be emphasized that the presence of elevated concentrations of chloride alone is not definitive proof of active saltwater intrusion, because chloride concentrations are naturally elevated near the boundary between freshwater and saltwater. Saltwater intrusion is indicated by increases in the chloride concentration of water samples collected periodically over time, rather than by a single concentration measured at one point in time (fig. 44).

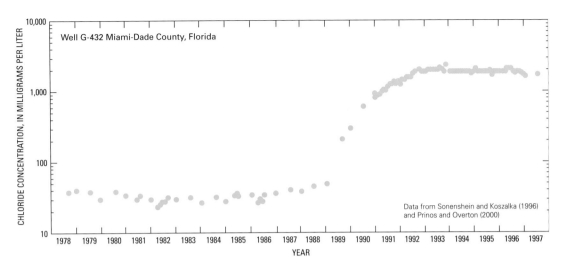

Figure 44. Chloride concentrations in water from the Biscayne aquifer near Biscayne Bay, Miami-Dade County, Florida. Chloride concentrations at the well remained low through 1988 but then increased sharply to about 2,000 milligrams per liter from 1989 to 1992. Movement of the saltwater interface in the Biscayne aquifer is affected by natural variations in recharge and evapotranspiration, ground-water withdrawals at supply wells, and drainage to and recharge from canals (Sonenshein and Koszalka, 1996). Monitoring of water levels and chloride concentrations in the Biscayne aquifer is being done collaboratively by Federal, State, and local agencies.

Special considerations should be given to the length and location of the screened (or open) interval of each monitoring well during the design of a saltwater-monitoring network. It is best if the wells are installed with short screens so that chloride concentrations and water levels accurately represent saltwater conditions for a very small volume of the aquifer (Reilly, 1993). Moreover, long-screened wells (or open intervals along a well bore) can induce mixing of waters of different chloride concentration because of vertical flow along the screened (or open) interval. Such mixing can lead to incorrect conclusions about the presence or the source of saltwater within an aquifer or multilayered aquifer system. Multiple wells arranged in triangular arrays and screened at the same horizontal plane can be used to estimate the direction and rate of saltwater movement within an aquifer.

Collecting a water sample from a saltwater-monitoring well in Fentress, Virginia. The sample is pumped from a depth of 759 feet below land surface from the Virginia Beach aquifer. Two additional monitoring wells are shown to the left of the well. Those wells monitor saltwater conditions in aquifers that lie above and below the Virginia Beach aquifer. The U.S. Geological Survey maintains a saltwater-monitoring network of about 105 wells in the Virginia Coastal Plain in collaboration with the Hampton Roads Planning District Commission and Virginia Department of Environmental Quality.

Robowell—An Automated Ground-Water Sampling Process to Remotely Monitor Saltwater Intrusion, Truro, Massachusetts

Monitoring of the freshwater-saltwater interface has been done primarily by manual collection of water samples from ground-water monitoring wells. Recently, an automated system called Robowell was developed to provide real-time measurement of changing ground-water levels and water-quality conditions at a monitoring location without the need for an onsite operator (Granato and Smith, 2002). The system consists of a submersible pump and several electronic instruments that control the frequency and volume of water pumped from a monitoring well (fig. G–1). The water is analyzed onsite by the Robowell instrumentation for properties such as specific conductance, pH, temperature, and dissolved oxygen. Results of the analyses are accessed immediately by phone or satellite links and posted on the internet.

Robowell has been used to monitor the movement of the freshwater-saltwater interface below the screened interval of one of the supply wells at a well field located 1,500 ft from the ocean in Truro, Massachusetts (figs. G–2 and G–3). The well field has had a history of high chloride concentrations caused by saltwater intrusion (LeBlanc and others, 1986), and pumping rates at the well field are monitored closely to prevent future intrusion. Ground water is pumped twice each day by the Robowell system from two monitoring wells located about 18 and 38 ft below the screened interval of the supply well.

Specific conductance of the ground water has been used as an indicator of saltwater beneath the supply wells because it is a direct measure of the

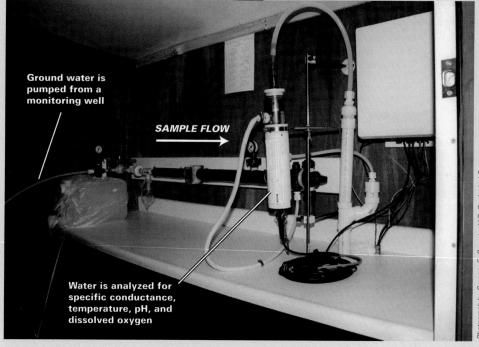

Figure G–1. The specific conductance of ground water pumped from a monitoring well at a public-supply well field in Truro, Massachusetts, is measured twice each day by use of an automated ground-water sampling system called Robowell. Some of the instrumentation for the system is shown in this photograph.

total dissolved-solids concentration of the water and is strongly correlated to the chloride concentrations of the water. Specific conductance measured at one of the two monitoring wells was relatively constant between February and late September 2001, ranging from about 17,000 to 19,000 microsiemens per centimeter (fig. G–3), which correlates to chloride concentrations of about 5,900 to 6,600 mg/L (G.E. Granato, U.S. Geological Survey, written commun., 2002). The total pumping rate at the well field during the same period ranged from less than 0.1 to almost 0.4 Mgal/d. However, on September 25, pumping from the well field abruptly ceased. As shown by the graph of specific conductance, continuous monitoring at the well provided a detailed history of the response of the freshwater-saltwater interface to the cessation of pumping. Specific conductance of the water dropped sharply during the 9-week shut-down period as the interface moved away from the well screen in response to the wells being removed from service. After the wells began pumping again in November, the specific conductance again increased as the interface moved back toward the supply well.

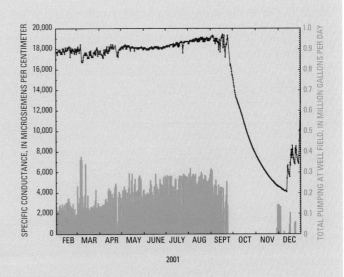

Figure G–3. Graph showing specific conductance of ground water pumped from one of the monitoring wells at the Truro well field during February through December 2001. On September 25, pumping at the well field abruptly ceased. In response, the specific conductance of water pumped from the monitoring well decreased sharply as the freshwater-saltwater interface moved away from the supply wells.

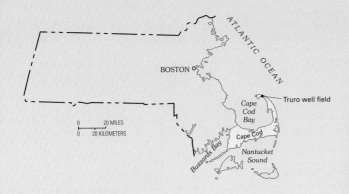

Figure G–2. Location of the well field in Truro, Massachusetts. In the background is Cape Cod Bay. The shelter in the foreground contains the Robowell system for monitoring saltwater intrusion.

Table 1. Saltwater-monitoring networks along the Atlantic coast

[Networks are arranged from north to south by State; no networks were identified for the States of Maine, New Hampshire, Massachusetts, Rhode Island, Connecticut, or Pennsylvania; USGS, U.S. Geological Survey]

State	Location	Agency(s)	Approximate number of monitoring wells in network	Frequency of sampling
New York	Long Island	USGS, New York City Department of Environmental Protection, Town of North Hempstead, Port Washington Water District	15	Each well sampled once each year
	Long Island (Nassau County)	Nassau County Department of Public Works	225	Variable: wells sampled every year, every other year, or every fourth year
	Long Island (Montauk area)	Suffolk County Water Authority	5	Wells sampled monthly to quarterly
New Jersey	Coastal areas of the State	New Jersey Department of Environmental Protection and USGS	[1]40	Each well sampled once each year
Delaware	New Castle County	Delaware Department of Natural Resources and Environmental Control	12	Each well sampled twice per year
	Sussex County	Delaware Geological Survey and Delaware Department of Natural Resources and Environmental Control	30	Variable—wells sampled twice annually, annually, or every other year
Maryland	Ocean City and Kent Island areas	USGS and Maryland Geological Survey	15	Each well sampled once per year
Virginia	Coastal Plain of entire State	USGS, Hampton Roads Planning District Commission, and Virginia Department of Environmental Quality	105	Ten wells sampled each year
North Carolina	Coastal Plain of entire State	North Carolina Department of Environment and Natural Resources (Division of Water Resources)	100	Each well sampled every 3 to 4 years
	Dare County/ Outer Banks	Dare County Water Department	30	Each well sampled biweekly
South Carolina	Beaufort and Colleton Counties	South Carolina Department of Health and Environmental Control	80	Forty wells monitored continuously; all wells sampled twice each year

Table 1. Saltwater-monitoring networks along the Atlantic coast—Continued

[Networks are arranged from north to south by State; no networks were identified for the States of Maine, New Hampshire, Massachusetts, Rhode Island, Connecticut, or Pennsylvania; USGS, U.S. Geological Survey]

State	Location	Agency(s)	Approximate number of monitoring wells in network	Frequency of sampling
Georgia	Savannah and Brunswick areas, Camden County	USGS, Georgia Department of Natural Resources (Environmental Protection Division), City of Brunswick, St. Johns River Water Management District (Florida)	90	Each well sampled once or twice each year
Florida[2]	Northeast Florida (primarily Duval County)	USGS and JEA	25	Each well sampled from 1 to 12 times each year
	Northeast Florida	St. Johns River Water Management District	300	Each well sampled quarterly to yearly
	East Orange County, City of Cocoa well field	USGS and City of Cocoa	40	Each well sampled from 1 to 12 times each year
	South Florida	USGS and several State and local agencies	100	Each well sampled from 2 to 12 times each year
	South Florida	South Florida Water Management District	[3]3,000	Each well sampled monthly to quarterly
	Southwestern Gulf counties	Southwest Florida Water Management District	345	Each well sampled from one to three times each year
	Sarasota County	USGS and Sarasota County	<5	Each well sampled semiannually
	Northcentral Counties	Suwannee River Water Management District	15	Each well sampled quarterly to yearly
	Northwestern Counties	Northwest Florida Water Management District	[3]200	Each well sampled quarterly to yearly

[1]In addition to the approximately 40 monitoring wells sampled each year by the USGS, the New Jersey Department of Environmental Protection mandates that purveyors collect water-quality samples quarterly from each of the State's large-capacity public-water supply wells. Because the chloride concentration of each water sample is one of the required constituents that is reported for each supply well, there is a large database of chloride concentrations available for public-supply wells in the State, in addition to those available for the monitoring wells sampled each year.

[2]Additional saltwater-monitoring networks are maintained by counties and cities that are not listed here; monitoring networks along the coast of the Gulf of Mexico are included for Florida.

[3]Water-supply and monitoring wells are sampled for chloride and other constituents as part of permitting requirements.

Use of Strontium Isotopes to Identify Sources of Saline Ground Water, Southwestern Florida

Prevention and management of saltwater intrusion often depend on understanding the origin of salinity in an aquifer or aquifer system. Saltwater intrusion in Lee and Collier Counties, southwestern Florida (fig. 45), is a major threat to fresh ground-water supplies for the two counties (Schmerge, 2001). Previous research has indicated that there are several paths by which saltwater might contaminate individual aquifers within the multilayered aquifer system, including lateral intrusion from the Gulf of Mexico and leakage between aquifers. To aid in the protection of the fresh ground-water resources of the two counties, the South Florida Water Management District and USGS recently completed a cooperative investigation to determine the extent and sources of salinity in the surficial and intermediate aquifer systems. As part of the study, naturally occurring isotopes of hydrogen, oxygen, and strontium were used in conjunction with chloride concentrations and the major-ion chemistry of individual water samples to determine the extent of seawater mixing and interaquifer leakage in the multilayered system. Isotopes are atoms of the same element that have different masses because of a difference in the number of neutrons in the atoms' nuclei. Hydrogen, for example, has three isotopes, ^1H, ^2H (deuterium), and ^3H (tritium), with relative masses of 1, 2, and 3, respectively. Naturally occurring isotopes have been widely used in ground-water studies to determine the source, flow path, and age of ground-water samples.

This case study focuses on the use of strontium isotopes to determine the origin of salinity in the lower Tamiami aquifer, which is the deepest aquifer in the surficial aquifer system and a major water-producing unit in the area. Chloride concentrations measured in the study area indicate a general trend of increasing salinity toward the coast and with depth, as illustrated by chloride concentrations mapped for the lower Tamiami aquifer that ranged from less than 100 mg/L in the northeastern section of the study area to greater than 5,000 mg/L toward the coast (fig. 46). Exceptions to this trend are an area near the coastline just north of Naples where relatively fresh water occurs seaward of more brackish water, and an extensive area southeast of Naples where the salinity of water in the aquifer is substantially higher than that in underlying aquifers.

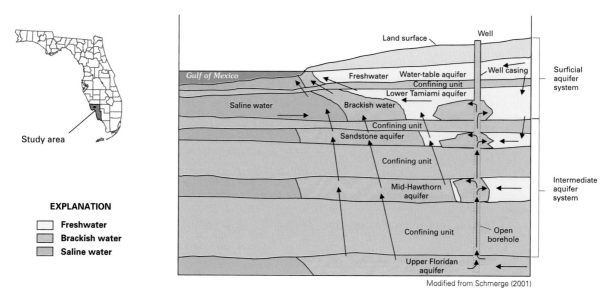

Figure 45. *Generalized ground-water flow patterns in coastal Lee and western Collier Counties, southwestern Florida. The figure illustrates some of the paths by which brackish and saline waters migrate in the multilayered system, including natural upward leakage between aquifers, brackish-water flow in a well with an open borehole, and lateral intrusion from the Gulf of Mexico.*

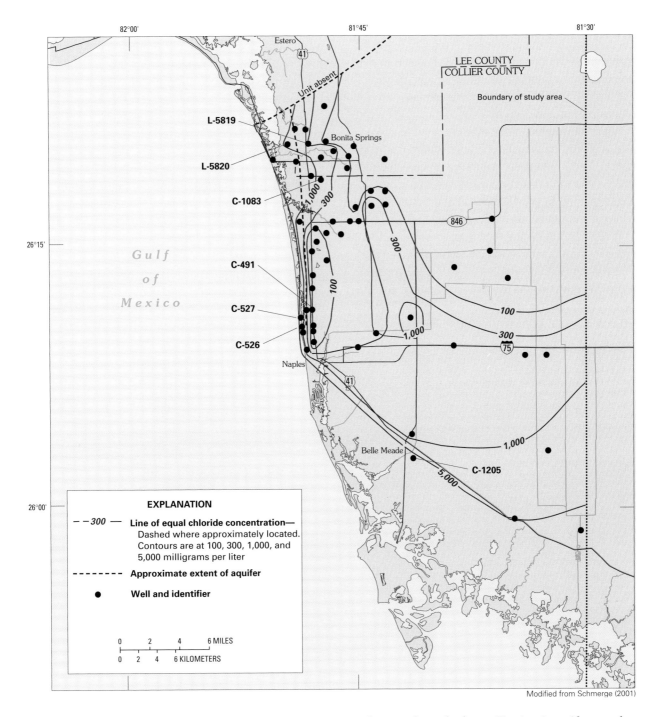

Figure 46. Lines of equal chloride concentration in ground water from the lower Tamiami aquifer, southwestern Florida.

Analysis of the major-ion chemistry of water samples from the lower Tamiami aquifer indicated a mixture of two different types of water—a freshwater characterized by concentrations of calcium and bicarbonate in excess of other major ions and a more saline water characterized by high concentrations of sodium and chloride. Moreover, water samples that had high chloride concentrations generally also had high strontium concentrations (fig. 47). Strontium concentrations in seawater are much greater than those in freshwater recharge; the linear trend shown on the graph between strontium and chloride concentrations is consistent with an interpretation of mixing between freshwater recharge and seawater.

The ratio of the concentrations of the strontium-87 and strontium-86 isotopes ($^{87}Sr/^{86}Sr$) can be used to determine the origin of dissolved strontium in a water sample. Because high strontium concentrations are associated with high chloride concentrations in the study area, the strontium isotopes in water samples could be used to determine the origin of salinity in the lower Tamiami aquifer (Schmerge, 2001). The basis of this approach is the fact that the $^{87}Sr/^{86}Sr$ ratio in seawater has varied over geologic time. For example, figure 48 shows that the $^{87}Sr/^{86}Sr$ ratio in seawater during the early Miocene ranged from about 0.7083 to 0.7087, but during the Oligocene was as low as about 0.7078. Most carbonate rocks and sediments that comprise the aquifers in southern Florida were formed in marine environments; as a consequence, the $^{87}Sr/^{86}Sr$ ratio of the rocks and sediments reflects the $^{87}Sr/^{86}Sr$ ratio of seawater during their formation. For example, a carbonate rock formed during the early Miocene would have a $^{87}Sr/^{86}Sr$ ratio ranging from about 0.7083–0.7087. As ground water flows through the rock, it will chemically equilibrate to a $^{87}Sr/^{86}Sr$ ratio that is equivalent to that of the rock. Therefore, ground waters that have come in contact with rocks of differing age will have contrasting strontium ratios.

Sediments that make up the lower Tamiami aquifer were deposited during the late Miocene to Pliocene. However, 12 of the 16 water samples collected from the aquifer had $^{87}Sr/^{86}Sr$ ratios that indicated that the water samples had come in contact with sediments that predated the late Miocene (fig. 48). Therefore, the lower Tamiami aquifer likely receives upward leakage from the underlying, and older, intermediate or Floridan aquifer systems (Schmerge, 2001). For example, water samples from wells C–1083, L–5820,

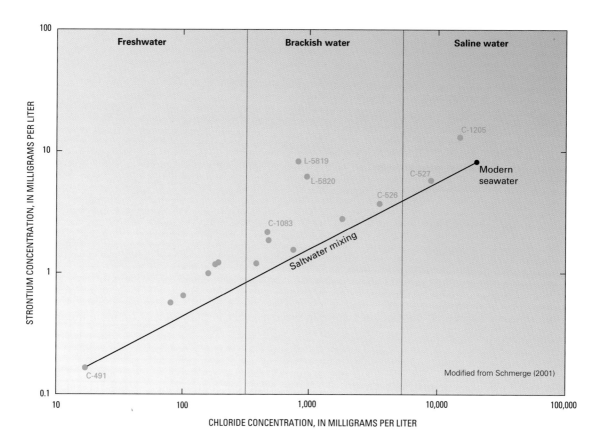

Figure 47. Relation between strontium and chloride concentrations in ground water from the lower Tamiami aquifer, southwestern Florida. The saltwater mixing line is drawn between the data point for water from well C–491, which is a freshwater well, and the data point for modern seawater.

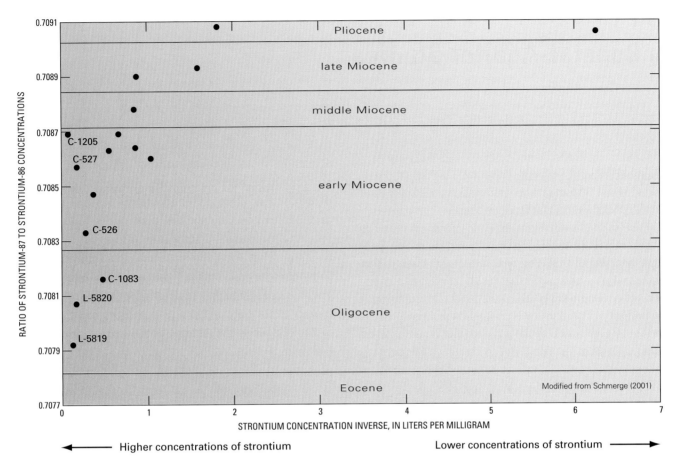

Figure 48. Relation between the ratio of strontium-87 to strontium-86 concentrations ($^{87}Sr/^{86}Sr$) and the inverse of strontium concentration in ground water from the lower Tamiami aquifer, southwestern Florida. Because inverse strontium concentrations were used to create the graph, water samples with the highest strontium concentrations plot on the left-hand side of the graph. The $^{87}Sr/^{86}Sr$ ratios of seawater during the different geologic epochs are indicated for reference. The approximate age range of each epoch (in millions of years before the present) is Pliocene, 2 to 5; Miocene, 5 to 24; Oligocene, 24 to 34; and Eocene, 34 to 55.

and L–5819 in the Bonita Springs area (fig. 46) plot within the Oligocene band in figure 48. The $^{87}Sr/^{86}Sr$ ratio of these three samples indicates that the underlying Upper Floridan aquifer is a likely source of water to the wells because carbonate rocks of the Upper Floridan aquifer are of Oligocene and older age. Underlying aquifers also were determined to be the primary source of salinity to the lower Tamiami aquifer because nearly all of the samples with strontium ratios indicative of older sediments corresponded with chloride concentrations greater than 300 mg/L.

The Gulf of Mexico is probably also a source of the saline water to the lower Tamiami aquifer along the coast; the gulf is the only known body of water in the study area with salinity concentrations that are high enough to account for the saline water that occurs in the aquifer (Schmerge, 2001). Chloride concentrations in water from wells C–526, C–527, and C–1205 (fig. 47) were 3,300, 8,300, and 14,000 mg/L, respectively. The chloride and strontium isotope data suggest that water from these three wells reflects mixing among freshwater recharge, upward leakage of brackish water, and saline water from the Gulf of Mexico.

Application of Geophysical Methods in Coastal-Aquifer Studies

Geophysical methods measure physical properties of the Earth that can be related to hydrologic or geologic aspects of an aquifer, such as pore-water conductivity (Stewart, 1999). Although there are a variety of geophysical techniques that commonly are applied in ground-water investigations, two types of techniques—electrical methods and seismic methods—are particularly useful in coastal environments. Electrical methods have been widely applied in coastal and island environments because of their ability to detect increases in the conductivity of an aquifer that result from increases in pore-water conductivity (Stewart, 1999). The electrical conductivity of an aquifer is controlled primarily by the amount of pore space of the aquifer (that is, the aquifer porosity) and by the salinity of the water in the pore space; increases in either the porosity or the concentration of dissolved ions result in increases in the conductivity of the ground water. Because seawater has a high concentration of dissolved ions, its presence in a coastal aquifer can be inferred from measurements of the spatial distribution of electrical conductivity. Seismic methods, on the other hand, do not detect saltwater, but can be used to delineate the distribution of geologic units within an aquifer that affect the distribution and movement of saltwater (Stewart, 1999). Three examples of the use of geophysical techniques to determine the distribution and movement of saltwater and to delineate paleochannels in aquifers along the Atlantic coast are described in the following section.

Electromagnetic Methods to Delineate Freshwater-Saltwater Interfaces and Saltwater Intrusion in South Florida

There are several surface and borehole geophysical methods that can be used to measure the conductivity of a ground-water system. One group of methods is based on the electromagnetic response of the Earth to electrical currents that are induced to flow in the ground. Surface electromagnetic techniques can be used to measure lateral and vertical spatial variations in the conductivity of an aquifer. Surface techniques are often the only practical way to collect conductivity data over large areas of an aquifer. Because the resolution with which a zone of saltwater can be detected by surface techniques decreases with increasing depth, borehole electromagnetic tools can be introduced directly into the subsurface to measure conductivity changes within an uncased borehole or a borehole lined with plastic casing (Stewart, 1999). Borehole techniques are particularly useful for delineating the transition zone from freshwater to saltwater at a monitoring well, and provide continuous measurement of aquifer conductivity that enhances water-quality samples collected at discrete sampling intervals at a well site.

Helicopter Electromagnetic Mapping of the Freshwater-Saltwater Interface, Everglades National Park

As described previously, the flow of freshwater into Everglades National Park has been severely diminished by the construction of an extensive system of canals that drain agricultural land south of Lake Okeechobee and that provide flood protection and drinking water for developed areas outside of the park. One of the tools that has been used to monitor hydrologic conditions in the Everglades is a helicopter-borne electromagnetic geophysical-surveying system that detects the extent of saltwater in the underlying Biscayne and surficial aquifers. The airborne surveys provide a means of rapidly and economically monitoring large areas of the Everglades where ground access is made difficult by standing water, sawgrass marshes, and mangrove swamps (Fitterman and Deszcz-Pan, 2001). Results of the surveys are being used to develop ground-water models of the area and to monitor restoration efforts.

The electromagnetic surveys make use of an instrument pod called a "bird" that is slung below a helicopter (fig. 49). The instrumentation consists of transmitter coils that induce electrical currents in the ground at different frequencies and receiver coils that measure the electromagnetic field produced by the induced currents. Maps of aquifer resistivity (the reciprocal of conductivity) produced at the different frequencies provide information about how resistivity varies with depth in the aquifer (Fitterman, 1996). The bird was flown during December 1994 at an altitude of about 100 ft above the ground over an area of 400 mi^2 along parallel lines about one-quarter mile apart.

Figure 49. Helicopter and equipment "bird" collecting electromagnetic geophysical data in Everglades National Park, Florida.

Two of the resistivity maps produced by the airborne survey demonstrate the general transition from areas of low conductivity (high resistivity) in the landward direction to areas of high conductivity (low resistivity) toward the shore (fig. 50). This increase in conductivity is interpreted to represent the transition between freshwater and saltwater in the aquifer. In the Shark River Slough area, the transition zone follows the terminus of tidally influenced rivers and streams.

The area underlying Taylor Slough, which is one of the primary sources of water to Florida Bay, is highly resistive (low conductivity), and the interface between freshwater and saltwater is much sharper than in the Shark River Slough area, perhaps because of a lack of tidally influenced streams in the Taylor Slough area (Fitterman and Deszcz-Pan, 1998). The area of low conductivity underlying Taylor Slough extends downward in the aquifer to at least 130 ft and may indicate that fresh surface

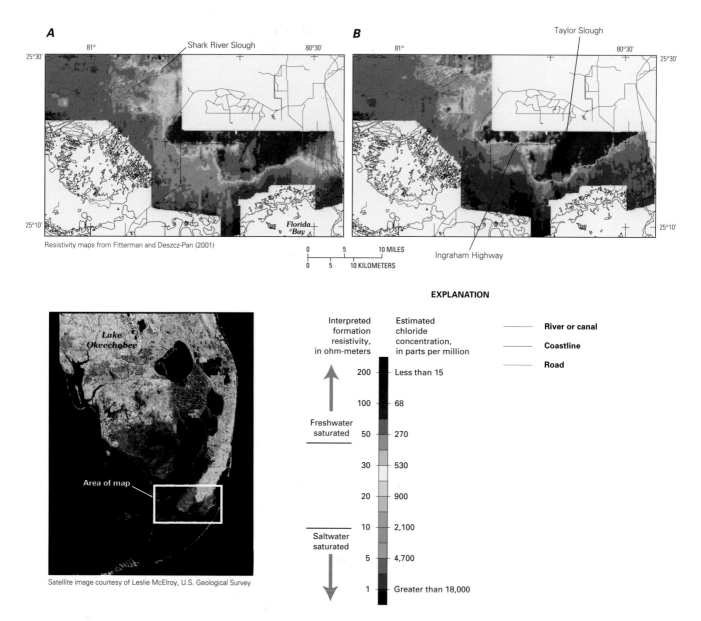

Figure 50. Satellite image of south Florida and resistivity maps of part of Everglades National Park. Interpreted resistivity maps represent depths in the ground-water system of about 15 feet (A) and 50 feet (B). The dark green and blue areas, which reflect relatively higher conductivity measurements, correspond to areas of saltwater in the Biscayne and surficial aquifers, whereas the red and brown areas correspond to freshwater parts of the aquifers.

water recharges the aquifer under the slough (Fitterman and Deszcz-Pan, 2001). The influence of manmade features can be seen by the area of relatively high conductivity along the Ingraham Highway (fig. 50B) where, in the past, saline water flowed inland in the canal adjacent to the highway. This area of high conductivity extends downward in the aquifer to depths of more than 50 ft, and is interpreted to reflect the strong hydraulic connection between flow in the canal and movement of the freshwater-saltwater interface in the Biscayne aquifer.

Results of the electromagnetic surveys have been supplemented with several ground-water-quality samples and additional land-based geophysical measurements to improve the horizontal and vertical resolution of the distribution of freshwater and saltwater in the aquifer. Water-quality samples also provide a means by which the resistivity results can be related to estimated chloride concentrations within the aquifer (fig. 50).

Borehole Electromagnetic Logs to Identify Saltwater Intrusion in an Urban Environment, Southeastern Florida

Rapid urban expansion in Palm Beach, Martin, and St. Lucie Counties, Florida (fig. 51), has increased the need for additional ground-water withdrawals from the shallow surficial aquifer system that is the primary source of drinking water for the three counties. The proximity of several of the area's well fields to the Atlantic coast has led to a concern that the increased freshwater demands may contribute to the intrusion of oceanic saltwater into the aquifer. Electromagnetic borehole-geophysical logs were used in conjunction with surface-geophysical methods and chloride-concentration data to map the position of the saltwater interface in the surficial aquifer system in the three counties during 1997–98 (Hittle, 1999). Borehole logs were made at 16 monitoring wells with a measurement tool that was lowered into each well.

The borehole electromagnetic logs provided detailed vertical profiles of the conductivity of the aquifer around each well. For example, data collected at well M–1289 near the Hobe Sound well field in eastern Martin County indicate zones of freshwater that lie above and below a zone of highly conductive saltwater present from about 85 to 140 ft below land surface (fig. 52A). Borehole-geophysical logs and chloride-concentration data collected near the well field indicate that a plug of saltwater is moving horizontally through a relatively permeable zone of the aquifer toward a production well located about one-half mile from Hobe Sound (fig. 52B). The permeable zone in which the saltwater is moving is composed of sandy limestone underlain by a less permeable unit of sand, limestone, and clay. The Hobe Sound well field area was the only location in the study area where freshwater was found below the farthest inland extent of the saltwater interface in the surficial aquifer.

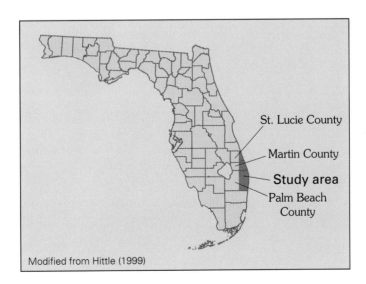

Figure 51. Area of eastern Palm Beach, Martin, and St. Lucie Counties, Florida, where geophysical and water-quality data were used to map the 1997–98 position of the freshwater-saltwater interface in the surficial aquifer.

Figure 52. (A) Borehole-geophysical log of the conductivity of the surficial aquifer at well M–1289 near the Hobe Sound well field, eastern Martin County, Florida. The zone of highly conductive saltwater at a depth of about 85 to 140 feet below land surface is overlain and underlain by freshwater. (B) Generalized freshwater-saltwater profile across the surficial aquifer near the Hobe Sound well field. Chloride concentrations greater than 100 milligrams per liter were considered to be evidence of saltwater mixing with freshwater in the aquifer and indicate the presence of the saltwater interface.

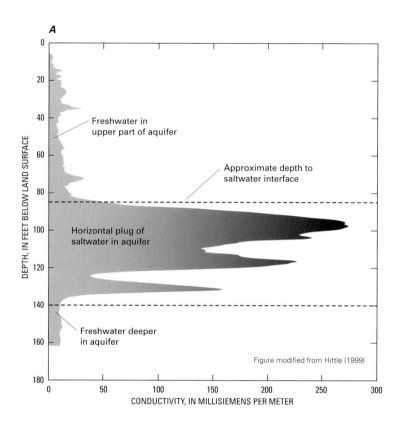

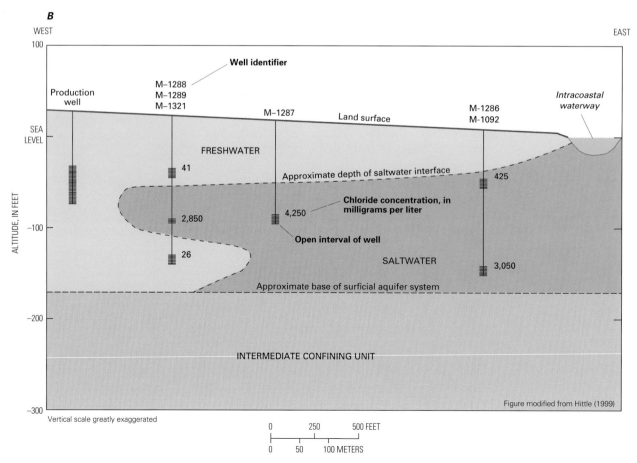

Seismic-Reflection Surveys to Delineate Paleochannels, North Carolina to Georgia

Seismic methods are based on variations in the mechanical strength and rigidity of geologic units that result from variations in compaction, cementation, and lithification (Stewart, 1999). One such method, seismic reflection, uses the reflection of sound waves off contacts between geologic units to map the location and geometry of the units (Stewart, 1999). Land-based and marine seismic-reflection surveys have been particularly useful in the delineation of paleochannels that formed along the Coastal Plain during periods of lowered sea level in the Pleistocene Epoch (about 2,000,000 to 10,000 years before present). During these periods, coastal rivers eroded underlying sediments as stream channels adjusted to lowered sea levels. Paleochannels are of concern where the rivers breached confining units and the erosional channels then filled with permeable materials as sea level rose again. In these situations, paleochannels can provide conduits for the movement of saltwater (or of land-based contaminants) into freshwater aquifers that underlie the confining units (fig. 53). Paleochannels have been identified as possible pathways for saltwater intrusion in northern Delaware along the Delaware River (Phillips, 1987); the northern Chesapeake Bay area, Maryland (Chapelle, 1985; Drummond, 1988; Phillips and Ryan, 1989); and the Port Royal Sound area, South Carolina (Landmeyer and Belval, 1996; Foyle and others, 2001; Krause and Clarke, 2001). The extent of the contribution of paleochannels to saltwater intrusion along the Atlantic coast is an area of active research.

Seismic-reflection surveys have been used to identify paleochannels in areas along the coast from North Carolina to Georgia. In Cherry Point, North Carolina, seismic-reflection surveys have been used with borehole geophysical logs and lithologic cores to identify paleochannels under the Neuse River estuary and adjoining land areas (fig. 54) (Daniel and others, 1996; Cardinell, 1999; Wrege and Daniel, 2001). In places, these paleochannels have cut through clayey confining units that separate freshwater aquifers in the area. The largest paleochannel ranges from 6,200 to 6,900 ft in width and extends beneath the Neuse River and the adjoining shoreline (Cardinell, 1999). Along the South Carolina and Georgia coastline, seismic-reflection surveys identified 11 areas where the confining unit overlying the Upper Floridan aquifer is thin or absent (Foyle and others, 2001). These areas, which cover an estimated 7 mi^2, are susceptible to seawater intrusion because of the absence of the confining unit and the presence of a downward hydraulic gradient from the ocean to the underlying Upper Floridan aquifer. Although most of the areas where the confining unit is absent are thought to have resulted from paleochannels, modern-day tidal scouring and dredging activities also may have contributed to the breaching of the confining unit along the coastline (Landmeyer and Belval, 1996; Foyle and others, 2001).

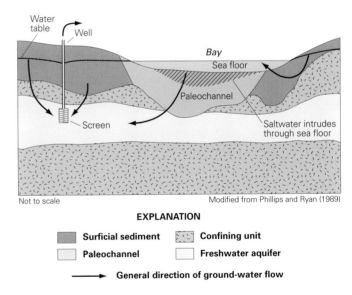

Figure 53. Preferential ground-water flow through a highly permeable paleochannel that breaches a confining unit and provides a conduit for saltwater to enter a freshwater aquifer.

Collecting seismic-reflection data on the Neuse River, Cherry Point, North Carolina.

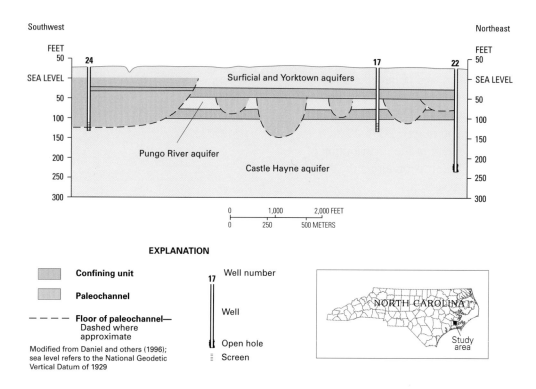

Figure 54. Representative hydrogeologic section showing paleochannels inferred from lithologic cores, seismic-reflection surveys, and other geophysical techniques, Cherry Point, North Carolina.

CHAPTER 4. GROUND WATER AND COASTAL ECOSYSTEMS

To this point, the focus of the report has been on the landward flow of saltwater into freshwater aquifers and the adverse effects that saltwater contamination can have on coastal ground-water supplies. Equally important, however, is the seaward flow of fresh ground water to coastal ecosystems and the role of ground water in delivering nutrients and other dissolved constituents to these systems. There are several ways in which ground water interacts with, and affects, coastal ecosystems (fig. 3). Ground-water discharge, or baseflow, to streams sustains the flow and aquatic habitats of coastal streams during periods when surface runoff is low. Ground-water discharge also helps to maintain water levels and water budgets of freshwater lakes, ponds, and wetlands. Dissolved chemical constituents discharged with ground water affect the salinity and geochemical budgets of coastal ecosystems and play a role in the biological species composition and productivity of these systems.

The role of ground water in delivering contaminants to coastal waters has become an area of growing interest and concern. Although coastal ground-water systems have been contaminated by many types of chemical constituents, much of the concern to date has focused on the discharge of excess nutrients, particularly nitrogen, to coastal ecosystems. Nutrient contamination of coastal ground water occurs as a consequence of activities such as wastewater disposal from septic systems and agricultural and urban uses of fertilizers. One of the most common effects of large inputs of nutrients to coastal waters is acceleration of the process of eutrophication, which is the enrichment of an ecosystem by organic material formed by primary productivity (that is, photosynthetic activity). Nutrient overenrichment can lead to excessive production of algal biomass, loss of important habitats such as seagrass beds and coral reefs, changes in marine biodiversity and distribution of species, and depletion of dissolved oxygen and associated die-offs of marine life (National Research Council, 2000).

Ground water reaches coastal environments either by direct discharge or as baseflow in the streams and rivers that drain coastal areas. Coastal water-resource management activities often have been directed toward riverine sources of contaminants. However, it is becoming evident that in some coastal settings direct ground-water discharge can be a substantial contributor of freshwater and dissolved constituents, particularly in coastal watersheds that consist of highly permeable soils that enable high rates of ground-water recharge and low rates of surface runoff. Because ground water moves slowly, the flushing of contaminated ground water from an aquifer can take many years, even several decades. To manage and protect coastal ecosystems, information is needed on the relative importance of ground water as a source of freshwater and contaminants in different types of coastal ecosystems.

In many coastal ecosystems, the hydrogeologic controls on ground-water discharge and contaminant loading are not well understood. This lack of knowledge results from the limited number of comprehensive studies of ground-water flow and geochemistry in different coastal settings and from the relative difficulty of measuring ground-water discharge and contaminant loading to coastal waters. Quantifying ground-water and contaminant discharge to coastal ecosystems and understanding the role of ground water in maintaining the biological health and geochemical balances of these systems increasingly require the integration of data-collection and data-analysis techniques from a diverse set of scientific fields. For example, hydrologists and oceanographers have independently developed techniques for measuring ground-water discharge rates to coastal waters, yet only recently have experiments to evaluate, compare, and improve the different measurement techniques been undertaken (Burnett and others, 2002). Innovative technologies for measuring and monitoring ground water in coastal environments continue to be developed and tested (Box H).

Three case studies are described to illustrate the role of ground water in different coastal ecosystems. The case studies move progressively seaward from a salt marsh, to a shallow coastal estuary, to a submarine spring located more than 2 mi from the Atlantic coastline. The contrasting modes of ground-water flow and discharge to the three coastal environments reflect large-scale differences in the geologic environments among the sites and smaller-scale heterogeneities that exist within the sediments and geologic formations at each site. Knowing the specific path by which ground water discharges to a coastal ecosystem is important because the natural removal of dissolved constituents from ground water depends on the sediments and geochemical conditions that the ground water encounters before discharging to the ecosystem.

New Technologies to Track Ground-Water Flow and Nutrient Transport to Coastal Bays of Delaware and Maryland

The Atlantic coastal bays of Delaware and Maryland receive significant but poorly quantified amounts of nutrients from ground-water discharge. These nutrients originate in agricultural, residential, and commercial areas within the watersheds of the bays. These coastal bays are representative of small estuarine lagoons behind barrier islands that are common along the Atlantic coast. Because of their restricted connection with the ocean, the bays typically have limited circulation, relatively long residence times, and high vulnerability to eutrophication (David Krantz, University of Toledo, Ohio, written commun., April 2002). Recent research has been done to track ground-water flow and nutrient transport to several of these bays (fig. H–1). A key component of the research has been the application of a suite of measurement tools to define the hydrogeologic framework near and beneath the bays and to map areas of ground-water discharge to the bays.

Innovative drilling and geophysical techniques have been used to map the sediments that make up the surficial aquifer and to determine the water chemistry and age of ground water beneath the bays. Two types of watercraft have provided platforms from which sediment cores, water-quality samples, and borehole geophysical logs could be collected. A construction barge was used for sample collection in the deeper waters of the bays where wind and tidal currents can create rough-water conditions (fig. H–2). Water samples and geophysical logs have been collected as much as 90 ft below the bay floor in core holes drilled from the barge. In shallow-water and tidal-flat areas, drilling has been done from the USGS Hoverprobe (fig. H–3), a hovercraft-mounted drilling rig that provides access to near-shore environments.

Electrical, direct-current "streamer" resistivity surveys were tested for their ability to detect fresh-water zones in sediments beneath the bays. The surveying equipment consists of a 360-ft electrical-cable system called a streamer that is towed behind a small vessel (fig. H–4). Initial surveys measured the electrical conductivity of bay sediments more than 100 ft beneath the seabed (fig. H–5). These surveys identified freshwater anomalies under the bays that are typically limited to several hundred feet from shore, but in some places extend more than one-half mile or more offshore (Manheim and others, 2001, 2002).

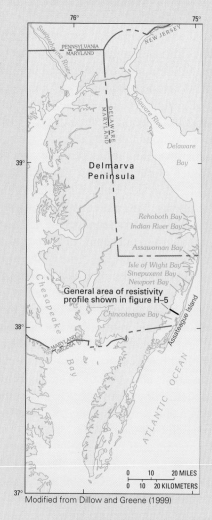

Figure H–1. Location of the Atlantic coastal bays of Delaware and Maryland, Delmarva Peninsula.

Figure H–2. *State of Delaware construction barge in Indian River Bay, Delaware, with drill rig mounted on the bow of the barge.*

Figure H–4. *Deployment of the resistivity-surveying equipment behind a small vessel in Indian River Bay, Delaware.*

Figure H–3. *Sediment coring and ground-water-quality sampling from the U.S. Geological Survey Hoverprobe in a tidal wetland of Maryland. Drilling is done by hydraulic vibracore equipment in the center of the hoverprobe craft.*

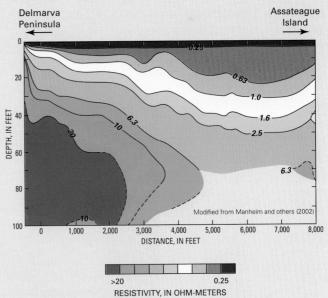

Figure H–5. *Representative resistivity profile across Chincoteague Bay, Maryland. The blue zones are interpreted to be fresh ground water flowing from the upland area west of the bay and mixing with saltwater beneath the bay (shown by the yellow to red zones). Location of resistivity profile shown on figure H–1.*

89

Ground-Water Discharge, Plant Distribution, and Nitrogen Uptake in a New England Salt Marsh

Salt marshes are important ecological and aesthetic resources along the Atlantic coast. Salt marshes provide fish and wildlife nurseries and habitat, form storm buffers, filter many waterborne contaminants, and are important to coastal food chains (Mitsch and Gosselink, 1993; Howes and others, 1996). Until quite recently, the role of salt marshes in intercepting coastal ground water and dissolved nutrients has been poorly understood. Such understanding is necessary, however, because of the importance of terrestrial inputs of nutrients to the eutrophication of nearshore waters. Recent investigations suggest that salt marshes adjacent to highly permeable aquifers intercept substantial amounts of ground-water flow, resulting in a significant reduction in the amount of upland nitrate reaching coastal waters (Howes and others, 1996).

One of the more intensively studied salt marshes along the Atlantic coast is the 200-acre Namskaket Marsh on Cape Cod, Massachusetts (fig. 55) (Weiskel and others, 1996; Howes and others, 1996; DeSimone and others, 1998). The marsh supports a diverse flora of grasses and flowering plants and abundant populations of invertebrates, fish, and birds. The inland part of Namskaket Marsh is adjacent to a facility that treats septage, which is the semisolid residue pumped from residential and commercial septic tanks. The treatment produces a nitrogen-rich effluent that is disposed of through infiltration into a highly permeable water-table aquifer (DeSimone and others, 1996). The effluent, which contains 40 to 50 mg/L of total nitrogen, has created a plume of contaminated ground water that is moving toward the marsh and Cape Cod Bay. Nitrogen within the contamination plume has undergone little removal during ground-water transport and is still largely mobile (DeSimone and Howes, 1996). Because nitrogen concentrations directly affect the productivity and species composition of salt-marsh plants and algae, increased nitrogen loading to the marsh through discharge of the plume has the potential to alter the distribution and productivity of wetland plants and algae and to indirectly affect wildlife in the marsh.

Aerial view of Namskaket Marsh and adjacent septage-treatment facility and infiltration beds (foreground). Cape Cod Bay is visible in the background.

Scientists have monitored the movement of the contamination plume since the beginning of effluent discharge in 1990. In addition, investigations have been made of the hydrologic, water-quality, biogeochemical, and ecological conditions of the marsh in advance of the arrival of the plume to better understand and predict possible responses of the marsh to nitrogen enrichment by contaminated ground water. Research in the marsh has focused on Inner Namskaket Marsh (fig. 55), a 17-acre area where discharge of contaminated ground water appears most likely. Most of the marsh is underlain by a silty-to-fibrous peat consisting of accumulations of organic matter that have relatively low hydraulic conductivity. The peat is typically 8 to 12 ft thick and has maximum accumulations of about 18 ft (Howes and others, 1996).

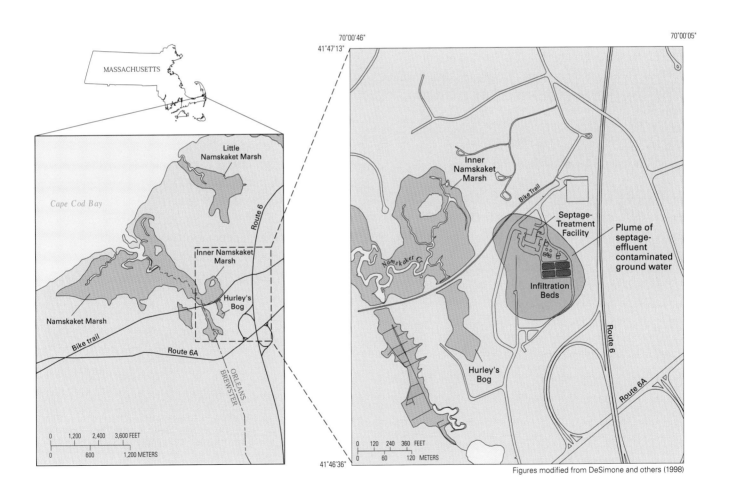

Figure 55. *Location of Namskaket Marsh and Creek, adjacent septage-treatment facility, and plume of septage-effluent contaminated ground water, Cape Cod, Massachusetts. The outline of the plume is based on water-quality data collected in November 1994.*

A conceptual understanding of ground-water flow and discharge to the marsh has been developed on the basis of detailed observations of the lithology and pore-water salinity of the marsh sediments, ground-water levels near and within the saturated marsh sediments, and seepage rates through the tidal creeks that drain the marsh. This conceptual understanding is illustrated by the ground-water flow paths and discharge zones to the marsh shown in figure 56. Ground-water discharge rates to the marsh are largest at the creek beds and at seepage zones near the boundary between the marsh and upland areas. In these two zones, the peat deposits are thin or absent, the hydraulic conductivity of the marsh sediments is greatest, and the upward hydraulic gradients between the underlying aquifer and marsh sediments are largest (Howes and others, 1996). Ground-water discharge within the interior vegetated areas of the marsh is comparatively small because of the thick deposits of peat, which have relatively low hydraulic conductivity and within which upward hydraulic gradients are relatively small. Some ground water flows under the marsh on deep flow paths and discharges directly to coastal waters.

The boundary seepage zones in the marsh are clearly illustrated by a map of pore-water salinity measured in the marsh sediments from 4 to 6 inches below the marsh surface (fig. 57). Inner Namskaket Marsh is flooded by tidal waters that have a salinity of about 17 parts per thousand (ppt). This salinity is significantly less than that of Cape Cod Bay (30 ppt), which is the source of the tidal water. The relatively low salinity of the creek waters is caused by fresh ground-water discharge to the marsh. The occurrence of nearly fresh water in the shallow marsh sediments periodically inundated by saltwater is an indication of ground-water discharge to the marsh. Areas of the marsh where pore-water salinity was less than 4 to 8 ppt corresponded to zones where persistent standing water or ground-water seepage was observed at the marsh surface (Weiskel and others, 1996). Pore-water salinities in the bottom sediments of the creek system also were generally less than 4 ppt, which is indicative of ground-water discharge.

The spatial distribution of the dominant plant species in Inner Namskaket Marsh broadly reflects the distribution of tidal flooding and ground-water discharge in the marsh (fig. 58) (DeSimone and others, 1998). As is typical of salt marshes along the Atlantic coast, the creek banks of the main channel, which are frequently flooded, are colonized by salt-tolerant cordgrass *(Spartina alterniflora)* except in the innermost brackish areas. Salt-marsh hay *(Spartina patens)* and spike grass *(Distichlis spicata)*, which are less tolerant of flooding by saline tidal waters, occur in areas primarily inland from creek banks. Common reed *(Phragmites australis)*, a fresh- to brackish-water plant, occurs primarily in irregular bands along the marsh/upland boundary where pore-water salinities of 4 to 8 ppt or less reflect the nearly continuous ground-water seepage in these areas. Other plants tolerant of brackish water also occur near marsh/upland boundaries and are shown with the areas of common reed in figure 58. Wrack deposits (dead plant material) and bare patches (panne) also are common. Physical disturbance to the marsh, such as from deposits of wrack, results in a dynamic and patchy distribution of plant species across the Inner Marsh.

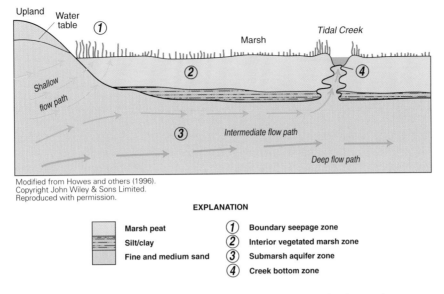

Figure 56. Diagram of ground-water flow paths and discharge locations below a tidal salt marsh in New England.

Ground-water seepage is visible on the marsh surface at the boundary between the marsh and upland areas.

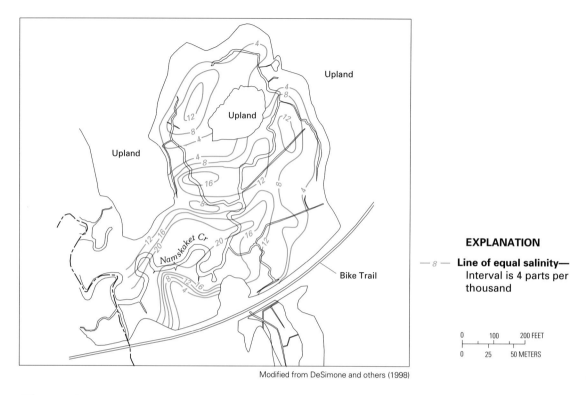

Figure 57. Pore-water salinity in peat sediment, Inner Namskaket Marsh, Cape Cod, Massachusetts, July 1995.

It is expected that if effluent-contaminated ground water discharges to the marsh, it will enter through the seepage zone at the marsh boundary or through the creek bottoms, rather than directly through the thick, low-permeability peat that underlies most of the marsh. Most of the nitrogen in the contamination plume is in the form of nitrate, which is a primary contributor to coastal eutrophication. As a consequence, there has been substantial interest in determining the potential for the marsh sediments to promote denitrification, which is the process by which subsurface bacteria transform dissolved nitrate to nitrogen gas in the absence of oxygen. Nitrogen gas is a harmless form of nitrogen that is naturally present in ground water and the atmosphere and does not cause eutrophication. The potential for denitrification is limited by the availability of organic carbon in the sediments through which ground water passes. The low organic-carbon concentrations of the aquifer sediments are the primary reason why denitrification does not remove large amounts of nitrate from the plume of contaminated ground water (DeSimone and Howes, 1996). However, experimental measurements of nitrate uptake in the creek-bottom and marsh sediments indicate that the higher organic-carbon content of these sediments probably will promote denitrification within the marsh and may remove substantial amounts of dissolved nitrate during ground-water discharge to the marsh (Weiskel and others, 1996).

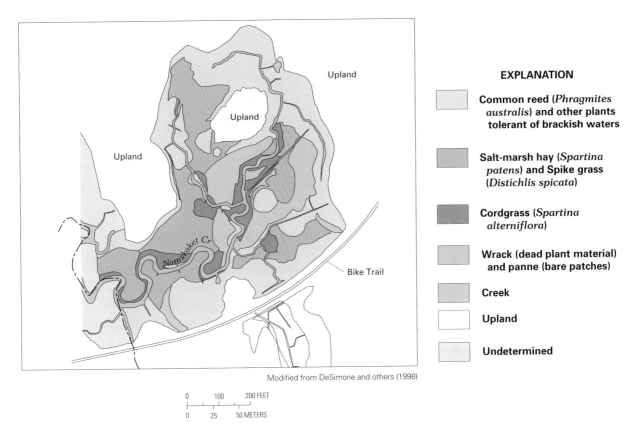

Figure 58. Distribution of wetland plants, Inner Namskaket Marsh, Cape Cod, Massachusetts, August through October 1995.

The banks of Namskaket Creek are bordered by salt-tolerant cordgrass (*Spartina altneriflora*), which in some areas is covered by dead plant material (wrack). Taller stands of common reed (*Phragmites australis*) are visible in front of the tree covered upland area.

Discharge of Nitrate-Contaminated Ground Water to a Coastal Estuary, Cape Cod, Massachusetts

The previous case study illustrated hydrogeologic controls on ground-water discharge in a coastal saltmarsh setting. The salt marsh intercepts ground water that otherwise would discharge directly into coastal waters. Interception of the ground water within the marsh has important implications for nitrogen loading of coastal waters because the sediments through which the ground water discharges in the salt marsh have relatively large amounts of organic matter that fuel microbial denitrification, which may reduce nitrate loading to coastal waters. A few miles east of the Namskaket Marsh site, ground water discharges directly to a shallow coastal estuary through subtidal sediments that do not provide opportunities for nitrogen removal because of their very low organic-matter content.

Nauset Marsh estuary is the most extensive and least-disturbed saltmarsh-estuary system within Cape Cod National Seashore, even though much of the estuary's watershed is developed for residential or commercial purposes (fig. 59) (Nowicki and others, 1999). Houses and businesses within the watershed are serviced by onsite sewage-disposal systems. Dissolved contaminants within the sewage, including nitrate and phosphate, seep downward to the aquifer and then are transported with the ground water toward the estuary. Although the estuary is fringed by a nearly continuous but narrow band of *Spartina alterniflora* marsh, sands and silts that have an organic content of less than 1 percent comprise 80 percent of the sediments that underlie the intertidal and subtidal areas of the estuary.

Three study sites were established along the shore of the estuary to determine spatial and temporal patterns of ground-water discharge and nitrogen loading to the estuary (fig. 59). The Salt Pond and Salt Pond Bay sites are downgradient from areas of low-housing density, and the Town Cove site is downgradient from areas of relatively high-housing density. Nitrate concentrations in ground water within the watershed of the Town Cove site averaged about 240 micromolar (about 3.4 mg/L), which was 40 times greater than average nitrate concentrations in ground water underlying the areas of low-housing density.

Detailed shoreline ground-water salinity surveys, hydraulic studies, and aerial thermal infrared imagery have shown that most fresh ground-water discharge to the estuary occurs in high-velocity seeps—areas of concentrated ground-water discharge—immediately seaward of the upland-fringing salt-marsh deposits. Moreover, ground-water discharge is highly variable along the shoreline, as shown by salinity and nitrate concentrations measured along the shore of Town Cove (fig. 60). Areas of substantial ground-water seepage along the Town

Northern end of Nauset Marsh, looking toward the Atlantic Ocean.

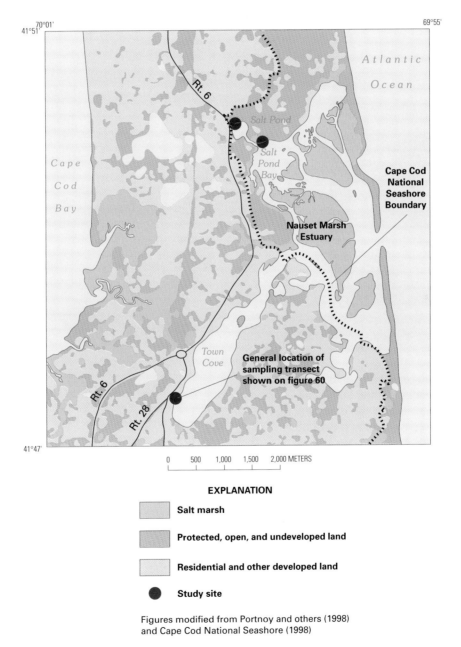

Figure 59. Nauset Marsh estuary and surrounding land uses.

Figure 60. Salinity and nitrate concentrations of ground-water samples collected in July 1993 along a sampling transect established near the mean, low-water line parallel to the shore of Town Cove (part of the Nauset Marsh estuary). Salinity of Nauset Marsh estuary ranges from 25 to 30 parts per thousand. (Note: a nitrate concentration of 100 micromolar is equivalent to a concentration of about 1.4 milligrams per liter). General location of transect shown in figure 59.

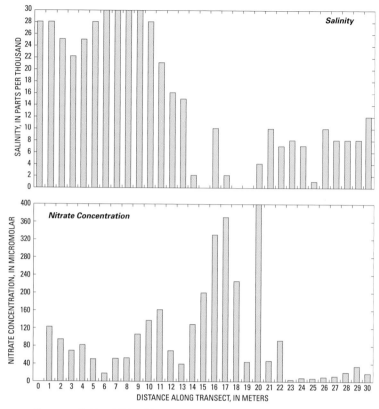

Figures modified from Urish and Qanbar (1997)

Seepage meters such as this one were used to measure ground-water discharge rates and the salinity and nitrate concentrations of ground-water discharge to Nauset Marsh estuary. Seepage meters consist of a bottomless chamber placed open-end down in the seabed. Water-quality samples are extracted through a port in the top of the chamber.

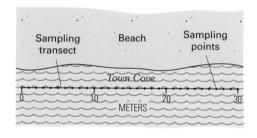

Sampling transect for ground-water quality. Sampling interval was 1 meter; samples were collected 10 centimeters below sediment surface.

Cove shoreline were evident at sampling points located from about 14 to 20 m along the transect, where salinity measurements were low and nitrate concentrations were high. The patchy nature of ground-water discharge along the shores of the estuary also was apparent in an aerial thermal image of Town Cove that identified areas of cold ground-water discharge (fig. 61) (Portnoy and others, 1998). The thermal infrared scan was done during low tide on a warm and clear summer evening when the contrast between cool, fresh ground water with a temperature of about 10 °C and relatively warm estuarine water with a temperature of about 22 °C was large. The cold-water plumes originated at the shore and spread seaward. Because the intruding ground water had a lower salinity than the estuarine water, the discharged ground water floated to the top of the estuary water column, which made the thermal contrast between the fresh ground water and the estuarine water most evident at the surface.

The temporal pattern of ground-water discharge to the estuary was closely related to the tidal cycle within the estuary (fig. 62) (Urish and Qanbar, 1997; Portnoy and others, 1998). During high tide when the tide level is greater than the underlying ground-water levels, saltwater infiltrates the beach slope and into the underlying aquifer (fig. 62A). As the tide recedes, water-level gradients on the beach slope are reversed—ground-water levels exceed the tide level, and fresh ground water and infiltrated saltwater begin to discharge (seep) onto the beach and directly into the estuary (fig. 62B). At the lowest tide, fresh ground water discharges directly onto the exposed beach near the low-tide line, although some of the infiltrated saltwater on the beach slope is retained in the intertidal beach zone landward of the low-tide line (fig. 62C). The cycle of saltwater infiltration and ground-water discharge is repeated with the next rising tide.

Nitrate concentrations in ground water discharging along the developed shores of Town Cove (where housing density is highest) were 10–50 times higher than those in ground water discharging along the shores of Salt Pond Bay, which is surrounded by the undeveloped lands of Cape Cod National Seashore. The coarse, sandy sediments through which the contaminated ground water rapidly discharges are well oxygenated and have very low organic content (Nowicki and others, 1999). The direct release of nitrate-contaminated ground water from these seeps into the estuarine waters circumvents the high denitrification potential of the poorly oxygenated, highly organic marsh and subtidal sediments. Because of the extent of exposed sediment at low tide below the fringe of saltmarsh, a large proportion of the ground-water discharge bypasses marsh peats, reducing the exchange between water and marsh material (Portnoy and others, 1998). As a consequence, a large fraction of the nitrogen dissolved in the ground water reaches the estuary, where it becomes available for plant and algal production.

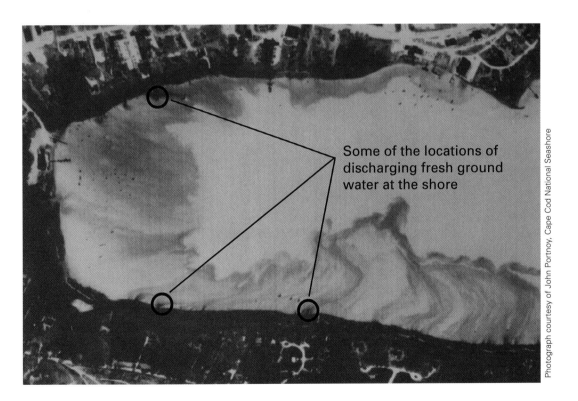

Figure 61. Aerial thermal infrared scan of Town Cove, Nauset Marsh. Discharging fresh ground water is visible as dark (relatively cold) streams flowing outward from shore over light-colored (warm) but higher density estuarine water. Data were collected at low tide at 9:00 p.m. eastern daylight time on August 7, 1994.

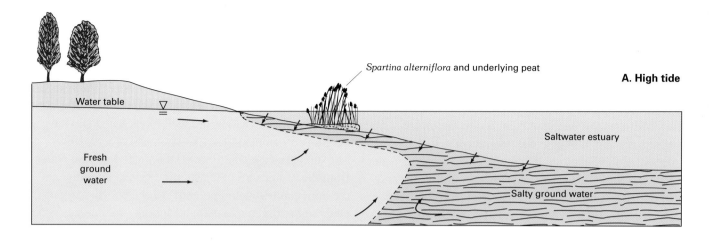
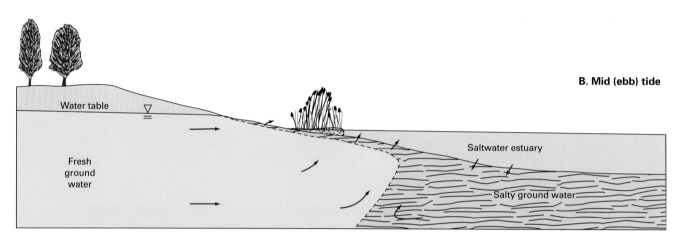
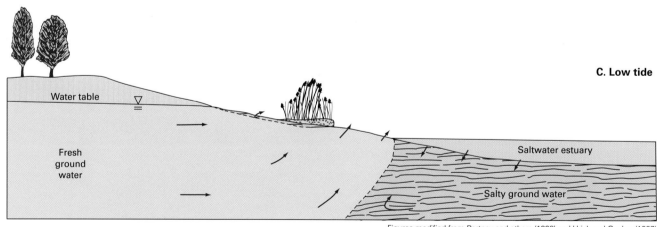

Figure 62. Ground-water discharge and saltwater infiltration at the aquifer-estuary boundary during a tidal cycle: (A) high tide; (B) mid (ebb) tide; and (C) low tide.

Submarine Ground-Water Discharge at Crescent Beach Spring, Florida

The previous case study illustrated diffuse ground-water seepage along the shoreline of a shallow coastal estuary underlain by sandy aquifer materials. A much different situation occurs off the coast of northeastern Florida, where a spectacular submarine spring discharges freshened ground water more than 2 mi from shore (fig. 63). Crescent Beach Spring appears to flow continuously through a single prominent vent in the Upper Floridan aquifer at rates that are consistent with a first-magnitude spring. The spring discharges from a large dissolution or collapse feature in the limestone aquifer (fig. 64) that appears to have been developed and maintained by the sustained submarine ground-water discharge (Swarzenski and others, 2001). The high flow rate of the spring creates a boil on the sea surface (fig. 65) and an odor of hydrogen sulfide that can be detected downwind of the discharge plume during calm sea conditions. Crescent Beach Spring is one of several submarine springs and sinks that have been identified along the Florida coast; the springs are most numerous along the State's gulf coast from the Tampa area northward and westward to the western panhandle (Rosenau and others, 1977).

Ground water discharged at the spring is thought to originally enter the aquifer through a network of lakes and sinkholes east of Gainesville, Florida (fig. 64), and the flow rate of the spring responds rapidly to changes in onshore precipitation and evaporation (Swarzenski and others, 2001). Fresh ground-water flow in the aquifer extends far beyond the shoreline of Crescent Beach because of the low permeability of the overlying clay-rich Hawthorn Formation, which forms a confining unit above the aquifer. At the spring, however, the Hawthorn and overlying younger deposits were eroded during a lowstand of the sea or were breached by a collapse or dissolution structure, and the aquifer is in direct contact with the Atlantic Ocean.

The vent of the spring extends from about 18 to 38 meters below sea level (fig. 66) and has a diameter of about 25 meters at its base. Water-quality sampling of the spring discharge has proven difficult because the depth of the spring throat is at the lower limit attainable by conventional SCUBA diving techniques and because of the extremely low visibility in the vent (Swarzenski and Reich, 2000). Scientists used a 1-meter-long well point driven into the shell hash and coarse sand in the deepest opening of the spring vent to collect spring discharge that had not been contaminated by seawater (fig. 66). The samples indicate that the spring water is substantially fresher and warmer than ambient seawater in the spring vent (fig. 67), which is consistent with an onshore source for the spring. Ongoing field investigations using a suite of naturally occurring geochemical tracers such as radium and strontium isotopes, methane, and radon gas are being conducted to better understand the sources of freshwater, ground-water flow paths, and ground-water traveltimes to the spring (Swarzenski and others, 2001).

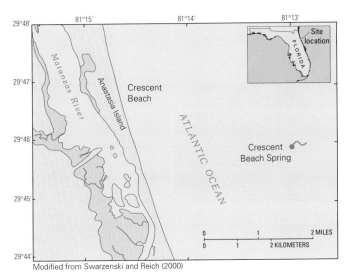

Figure 63. Location of Crescent Beach Spring, northeastern Florida.

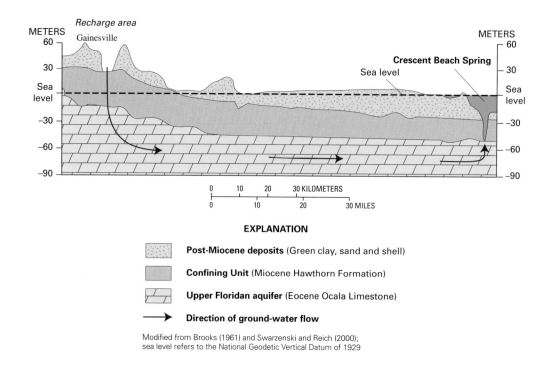

Figure 64. Idealized cross section of ground-water flow to Crescent Beach Spring.

Figure 65. Orange buoy marking the center of Crescent Beach Spring discharge plume. The edge of the boil is faintly visible.

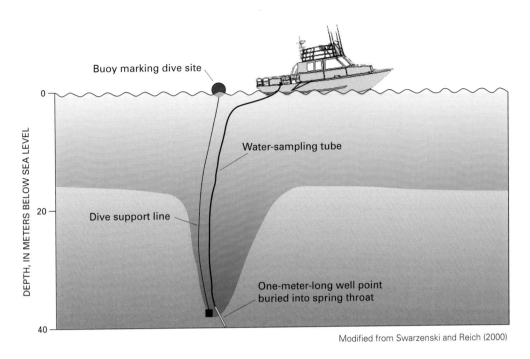

Figure 66. Sampling method for collecting submarine ground-water discharge at Crescent Beach Spring.

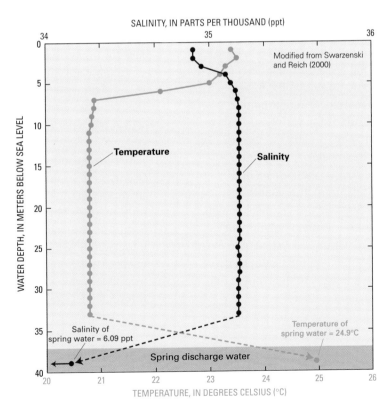

Figure 67. Profile of water salinity and temperature with depth down into the opening of the Crescent Beach Spring vent. Note extreme differences in temperature and salinity between spring water collected with the well point and ambient seawater above 35 meters depth.

CHALLENGES AND OPPORTUNITIES

The focus of this report has been to describe the current (2003) understanding of ground water in freshwater-saltwater environments of the Atlantic coast. As has been illustrated by the several case studies, much progress has been made in understanding the natural distribution of saltwater in aquifers of the region and the geologic, hydrologic, and geochemical processes that control freshwater and saltwater flow and mixing in coastal aquifers. The case studies highlight the critical role of geologic processes in determining the composition and structure of coastal aquifers and the short-term and long-term dynamic forces that drive the flow of freshwater and saltwater within these aquifers. The case studies also demonstrate that in many aquifers along the Atlantic coast, the natural occurrence and movement of saltwater has been changed by the development of ground-water resources for human uses. Ground-water development has lowered water levels and caused saltwater to intrude into many of the region's most productive aquifers. In response, substantial effort has been directed toward understanding the several pathways and processes that control saltwater movement through freshwater aquifers and the development of water-quality observation-well networks and other monitoring approaches to detect and track saltwater movement. Scientific monitoring and analysis have demonstrated that the different modes of saltwater intrusion are the result of the variety of hydrogeologic settings, sources of saline water, and history of ground-water pumping and freshwater drainage along the Atlantic coast, and that effective monitoring and management of saltwater intrusion require that these factors be known. Less well understood are the mechanisms and rates of ground-water discharge and contaminant loading to coastal ecosystems. These ecosystems are of high intrinsic and economic value, yet the health of these systems can be threatened by changes in the amount and quality of water that is discharged to them.

Recent population increases along the Atlantic coastal zone suggest that demands on the ground-water resources of the region will grow in the coming years. The need for water to support coastal populations and economic prosperity will present scientists, water-resource managers, and public decisionmakers with a number of challenges and opportunities for understanding and wisely managing coastal ground-water resources. Hydrologic studies and data-collection activities, as they have in the past, will contribute to the development and management of coastal ground-water resources. However, there are a number of scientific issues related to ground water in freshwater-saltwater environments that will need to be addressed. Some of these issues are described in the following paragraphs:

- *Periodic evaluation of the adequacy of ground-water monitoring networks and estimates of ground-water use*: Observation-well networks that monitor ground-water levels and ground-water quality are indispensable for determining the effects of ground-water development on ground-water levels and ground-water storage and for monitoring the location and movement of saline water in coastal aquifers. Although many such networks are already in place along the Atlantic coast, recent increases in population will require periodic evaluation of the adequacy of these networks to monitor changes in ground-water levels and the movement of saline ground water. Such evaluation is particularly important in areas where ground-water development has only recently begun or is substantially increasing. Similarly, water-use estimates are vital for understanding human impact on water and ecological resources and for assessing whether available water supplies will be adequate to meet future needs. Commonly, however, estimates of ground-water use are not available to identify trends in water use and the potential for overuse of the resource.

- *Improved understanding of the controls on saltwater occurrence and intrusion*: Much work has been done to understand the geologic, hydrologic, and geochemical controls on saltwater occurrence and intrusion in freshwater aquifers along the Atlantic coast, particularly where saltwater intrusion is a recognized problem. Moreover, this report has described several locations along the Atlantic coast where the sources of saltwater and the modes of saltwater intrusion are being actively researched—examples include the role of the Chesapeake Bay impact crater on Virginia's inland saltwater wedge, the effects of paleochannels on saltwater intrusion along the Georgia and Carolina coasts, and the role of fractures and other structural anomalies on saline-water movement in northeastern Florida. As coastal populations expand into previously undeveloped areas, it is likely that new patterns of saltwater occurrence and intrusion will be identified that will require scientific analysis and monitoring.

- *The role of ground water in coastal ecosystems*: There is a continuing need to quantify the relative importance of ground water as a source of freshwater and contaminants to different types of coastal ecosystems. As described in this report, coastal water-resource management activities often have been directed toward riverine sources of contaminants. A better understanding is needed of the hydrogeologic and geochemical controls on the rates, locations, and quality of ground-water discharge to coastal ecosystems that receive substantial ground-water discharge in comparison to other sources of discharge. Understanding ground-water contributions to coastal ecosystems will require an integrated effort from the fields of hydrology, geology, geochemistry, and biology. A related need is to quantify the residence times and traveltimes of ground water and associated contaminants in coastal aquifers. Because ground water moves slowly, the flushing of contaminated ground water from coastal aquifers to receiving wetlands and surface waters can take many years or decades.

- *Scientific evaluations to support ground-water management*: There are several areas in which scientific evaluations are needed to support traditional and emerging approaches for ground-water management in coastal regions. As water use increases, communities are beginning to look for more effective ways to conjunctively use ground- and surface-water resources and to seek alternative ground-water supplies from aquifers that have not been previously used. The use of desalination and aquifer storage and recovery systems is becoming more widespread in coastal regions, and their use is likely to increase in the future. Scientific studies can contribute to the success of desalination systems by identifying brackish and saline ground-water resources that are appropriate for desalination and by characterizing the hydrogeologic and geochemical conditions of the aquifers that contain the brackish and saline waters. Similarly, scientific evaluations can be conducted to characterize the hydrogeologic and biogeochemical controls that affect the efficiency of aquifer storage and recovery systems. New developments in ground-water flow and transport modeling and in computer-visualization techniques will increase the ability of hydrologists to simulate the effects of proposed ground-water development scenarios on ground-water and surface-water resources and on the paths and rates of saltwater intrusion, and to better communicate the results of these simulations to water-resource managers.

ACKNOWLEDGMENTS

The author thanks several colleagues for their input to and review of this report.[1] William M. Alley, John F. Bratton, John S. Clarke, Gregory S. Gohn, Anthony S. Navoy, Thomas E. Reilly, Charles T. Roman (National Park Service), S. Jeffress Williams, and Carl Zimmerman (National Park Service) reviewed the draft report in its entirety. Several people provided information for, or review of, the case studies described in the report: Leslie A. DeSimone, David V. Fitterman, Gregory E. Granato, Clinton D. Hittle, B. Barbara Howie, David E. Krantz (University of Toledo), Pierre J. Lacombe, Christian D. Langevin, Curtis E. Larsen, Denis R. LeBlanc, Frank T. Manheim, E. Randolph McFarland, Patricia A. Metz, Martha G. Nielsen, John W. Portnoy (National Park Service), C. Wylie Poag, Ronald S. Reese, Robert A. Renken, Ward E. Sanford, David L. Schmerge, Roy S. Sonenshein, Rick M. Spechler, Mark T. Stewart (University of South Florida), Kevin J. Stover (Southwest Florida Water Management District), Peter W. Swarzenski, Dorothy H. Tepper, E. Robert Thieler, Peter K. Weiskel, and Beth M. Wrege. Illustrations for the report were prepared by Margo J. VanAlstine, John M. Evans, and Lance J. Ostiguy. Carol L. Anderson edited the manuscript, and Joy K. Monson prepared the final manuscript.

REFERENCES CITED

Alley, W.M., 2003, Desalination of ground water—Earth science perspectives: U.S. Geological Survey Fact Sheet FS–075–03, 4 p.

Anderson, M.P., and Woessner, W.W., 1992, Applied ground-water modeling—Simulation of flow and advective transport: San Diego, Calif., Academic Press, 381 p.

Ayers, M.A., Wolock, D.M., McCabe, G.J., Hay, L.E., and Tasker, G.D., 1994, Sensitivity of water resources in the Delaware River Basin to climate variability and change: U.S. Geological Survey Water-Supply Paper 2422, 42 p.

Back, William, and Freeze, R.A., eds., 1983, Chemical hydrogeology: Benchmark Papers in Geology, 73, Hutchinson Ross Publication Company, Stroudsburg, Pa., 416 p.

Barlow, P.M., and Wild, E.C., 2002, Bibliography on the occurrence and intrusion of saltwater in aquifers along the Atlantic coast of the United States: U.S. Geological Survey Open-File Report 02–235, 30 p.

Basu, A.R., Jacobsen, S.B., Poreda, R.J., Dowling, C.B., and Aggarwal, P.K., 2001, Large groundwater strontium flux to the oceans from the Bengal Basin and the marine strontium isotope record: Science, v. 293, p. 1470–1473.

Bear, Jacob, Cheng, A.H.-D., Sorek, Shaul, Ouazar, Driss, and Herrera, Ismael, eds., 1999, Seawater intrusion in coastal aquifers—Concepts, methods and practices: Dordrecht, The Netherlands, Kluwer Academic Publishers, 625 p.

Blair, D.A., Spronz, W.D., and Ryan, K.W., 1999, Brackish groundwater desalination—A community's solution to water supply and aquifer protection: Journal of the American Water Resources Association, v. 35, no. 5, p. 1201–1212.

Brooks, H.K., 1961, The submarine spring off Crescent Beach, Florida: Quarterly Journal of the Florida Academy of Sciences, v. 24, no. 2, p. 122–134.

Burnett, Bill, Chanton, Jeff, Christoff, Jamie, Kontar, Evgeny, Krupa, Steve, Lambert, Michael, Moore, Willard, O'Rourke, Daniel, Paulsen, Ronald, Smith, Christopher, Smith, Leslie, and Taniguchi, Makoto, 2002, Assessing methodologies for measuring groundwater discharge to the ocean: Eos, v. 83, no. 11, p. 117–123.

Buros, O.K., 2000, The ABCs of desalting (2d ed.): Topsfield, Mass., International Desalination Association, 30 p.

Buxton, H.T., and Shernoff, P.K., 1999, Ground-water resources of Kings and Queens Counties, Long Island, New York: U.S. Geological Survey Water-Supply Paper 2498, 113 p.

---

[1]Affiliation is U.S. Geological Survey, unless otherwise noted.

Cape Cod National Seashore, 1998, Forging a collaborative future—Final environmental impact statement for the general management plan: U.S. National Park Service, Report NPS D–137A, v. 1, 374 p.

Cardinell, A.P., 1999, Application of continuous seismic-reflection techniques to delineate paleochannels beneath the Neuse River at U.S. Marine Corps Air Station, Cherry Point, North Carolina: U.S. Geological Survey Water-Resources Investigations Report 99–4099, 29 p.

Caswell, W.B., 1979a, Maine's ground-water situation: Ground Water, v. 17, no. 3, p. 235–243.

Caswell, W.B., 1979b, Ground water handbook for the State of Maine: Maine Geological Survey, 126 p.

Caswell, W.B., 1987, Ground water handbook for the State of Maine (2d ed.): Maine Geological Survey Bulletin 39, 135 p.

Chapelle, F.H., 1985, Hydrogeology, digital solute-transport simulation, and geochemistry of the lower Cretaceous aquifer system near Baltimore, Maryland, *with a section on* Well records, pumpage information, and other supplemental data, by T.M. Kean: Maryland Geological Survey Report of Investigations 43, 120 p.

Church, T.M., 1996, An underground route for the water cycle: Nature, v. 380, no. 6575, p. 579–580.

Cooper, H.H., Jr., 1964, A hypothesis concerning the dynamic balance of fresh water and salt water in a coastal aquifer: U.S. Geological Survey Water-Supply Paper 1613–C, p. 1–12.

Custodio, E., 1997, Detection, chap. 2 *in* Seawater intrusion in coastal aquifers—Guidelines for study, monitoring and control: Rome, Italy, Food and Agriculture Organization of the United Nations Water Reports 11, p. 6–22.

Daniel, C.C., III, Miller, R.D., and Wrege, B.M., 1996, Application of geophysical methods to the delineation of paleochannels and missing confining units above the Castle Hayne aquifer at U.S. Marine Corps Air Station, Cherry Point, North Carolina: U.S. Geological Survey Water-Resources Investigations Report 95–4252, 106 p.

Delaware River Basin Commission, 2001, The salt line—What is it and where is it?: Delaware River Basin Commission, accessed March 8, 2001, at URL *http://www.state.nj.us/drbc/salt.html*

DeSimone, L.A., Barlow, P.M., and Howes, B.L., 1996, A nitrogen-rich septage-effluent plume in a glacial aquifer, Cape Cod, Massachusetts, February 1990 through December 1992: U.S. Geological Survey Water-Supply Paper 2456, 89 p.

DeSimone, L.A., and Howes, B.L., 1996, Denitrification and nitrogen transport in a coastal aquifer receiving wastewater discharge: Environmental Science and Technology, v. 30, no. 4, p. 1152–1162.

DeSimone, L.A., Howes, B.L., Goehringer, D.G., Weiskel, P.K., 1998, Wetland plants and algae in a coastal marsh, Orleans, Cape Cod, Massachusetts: U.S. Geological Survey Water-Resources Investigations Report 98–4011, 33 p.

Dillow, J.J.A., and Greene, E.A., 1999, Ground-water discharge and nitrate loadings to the coastal bays of Maryland: U.S. Geological Survey Water-Resources Investigations Report 99–4167, 8 p.

Douglas, B.C., Kearney, M.S., and Leatherman, S.P., 2001, Sea level rise—History and consequences: San Diego, Calif., Academic Press, 232 p.

Drummond, D.D., 1988, Hydrogeology, brackish-water occurrence, and simulation of flow and brackish-water movement in the Aquia aquifer in the Kent Island area, Maryland: Maryland Geological Survey Report of Investigations 51, 131 p.

Fairbanks, R.G., 1989, A 17,000-year glacio-eustatic sea level record—Influence of glacial melting rates on the Younger Dryas event and deep-ocean circulation: Nature, v. 342, p. 637–642.

Feth, J.H., 1981, Chloride in natural continental water—A review: U.S. Geological Survey Water-Supply Paper 2176, 30 p.

Feth, J.H., and others, 1965, Preliminary map of the conterminous United States showing depth to and quality of shallowest ground water containing more than 1,000 parts per million dissolved solids: U.S. Geological Survey Hydrologic Investigations Atlas 199, 31 p.

Fies, M.W., Renken, R.A., and Komlos, S.B., 2002, Considerations for regional ASR in restoring the Florida Everglades, USA, *in* Dillon, P.J., Management of aquifer recharge for sustainability, Proceedings of the 4th International Symposium on Artifical Recharge, Adelaide, Australia, September 22–26, 2002: The Netherlands, A.A. Balkema Publishers, p. 341–346.

Fitterman, D.V., 1996, Geophysical mapping of the freshwater/saltwater interface in Everglades National Park, Florida: U.S. Geological Survey Fact Sheet 173–96, 2 p.

Fitterman, D.V., and Deszcz-Pan, Maryla, 1998, Helicopter EM mapping of saltwater intrusion in Everglades National Park, Florida: Exploration Geophysics, v. 29, p. 240–243.

Fitterman, D.V., and Deszcz-Pan, Maryla, 1999, Geophysical mapping of saltwater intrusion in Everglades National Park, *in* 3rd International Symposium on Ecohydraulics, Salt Lake City, Utah, July 12–16, 1999, p. 18 (on CD-ROM).

Fitterman, D.V., and Deszcz-Pan, Maryla, 2001, Using airborne and ground electromagnetic data to map hydrologic features in Everglades National Park, *in* Symposium on the Application of Geophysics to Engineering and Environmental Problems (SAGEEP 2001), Denver, Colo., March 4–7, 2001, Proceedings: Denver, Colo., Environmental and Engineering Geophysical Society, 17 p. (on CD-ROM).

Foyle, A.M., Henry, V.J., and Alexander, C.R., 2001, The Miocene aquitard and the Floridan aquifer of the Georgia/South Carolina coast—Geophysical mapping of potential seawater intrusion sites: Georgia Geologic Survey Bulletin 132, 61 p.

Freeze, R.A., and Cherry, J.A., 1979, Groundwater: Englewood Cliffs, N.J., Prentice-Hall, 604 p.

Georgia Environmental Protection Division, 1997, Interim strategy for managing saltwater intrusion in the Upper Floridan aquifer of southeast Georgia, April 23, 1997: Atlanta, Georgia Environmental Protection Division, 19 p.

Gill, H.E., 1962, Ground-water resources of Cape May County, New Jersey—Salt-water invasion of principal aquifers: New Jersey Department of Conservation and Economic Development, Division of Water Policy and Supply Special Report 18, 171 p.

Gohn, G.S., Bruce, T.S., Catchings, R.D., Emry, S.R., Johnson, G.H., Levine, J.S., McFarland, E.R., Poag, C.W., and Powars, D.S., 2001, Integrated geologic, hydrologic, and geophysical investigations of the Chesapeake Bay impact structure, Virginia, USA—A multi-agency program [abs.]: 32nd Lunar and Planetary Science Conference, March 12–16, 2001, Houston, Texas, Abstract 1901.

Granato, G.E., and Smith, K.P., 2002, Robowell—Providing accurate and current water-level and water-quality data in real time for protecting ground-water resources: U.S. Geological Survey Fact Sheet FS–053–02, 6 p.

Guo, Weixing, and Bennett, G.D., 1998, Simulation of saline/fresh water flows using MODFLOW, in Poeter, E., and others, MODFLOW '98 Conference, Golden, Colo., 1998, Proceedings: Golden, Colo., v. 1, p. 267–274.

Guo, Weixing, and Langevin, C.D., 2002, User's guide to SEAWAT—A computer program for simulation of three-dimensional variable-density ground-water flow: U.S. Geological Survey Techniques of Water-Resources Investigations, book 6, chap. A7, 77 p.

Heath, R.C., 1984, Ground-water regions of the United States: U.S. Geological Survey Water-Supply Paper 2242, 78 p.

Heath, R.C., 1998, Basic ground-water hydrology: U.S. Geological Survey Water-Supply Paper 2220 (eighth printing), 86 p.

Hem, J.D., 1989, Study and interpretation of the chemical characteristics of natural water (3d ed.): U.S. Geological Survey Water-Supply Paper 2254, 263 p.

Hittle, C.D., 1999, Delineation of saltwater intrusion in the surficial aquifer system in eastern Palm Beach, Martin, and St. Lucie Counties, Florida, 1997–98: U.S. Geological Survey Water-Resources Investigations Report 99–4214, 1 sheet.

Howes, B.L., Weiskel, P.K., Goehringer, D.D., and Teal, J.M., 1996, Interception of freshwater and nitrogen transport from uplands to coastal waters—The role of saltmarshes, in Nordstrom, K.F., and Roman, C.T., eds., Estuarine shores—Evolution, environments and human alterations: New York, John Wiley, p. 287–310.

Hughes, J.L., 1979, Saltwater barrier line in Florida—Concepts, considerations, and site examples: U.S. Geological Survey Water-Resources Investigations 79–75, 29 p.

Intergovernmental Panel on Climate Change, 2001, Intergovernmental Panel on Climate Change Third Assessment Report—Climate change 2001, synthesis report: Intergovernmental Panel on Climate Change, 111 p., accessed November 4, 2002, at URL *http://www.ipcc.ch/pub/reports.html*

Johannes, R.E., 1980, The ecological significance of the submarine discharge of groundwater: Marine Ecology Progress Series, v. 3, p. 365–373.

Johnston, R.H., and Bush, P.W., 1988, Summary of the hydrology of the Floridan aquifer system in Florida and in parts of Georgia, South Carolina, and Alabama: U.S. Geological Survey Professional Paper 1403–A, 24 p.

Johnston, R.H., Bush, P.W., Krause, R.E., Miller, J.A., and Sprinkle, C.L., 1982, Summary of hydrologic testing in Tertiary limestone aquifer, Tenneco offshore exploration well—Atlantic OCS, Lease-block 427 (Jacksonville NH 17–5): U.S. Geological Survey Water-Supply Paper 2180, 15 p.

Jones, B.F., Vengosh, Avner, Rosenthal, Eliyahu, and Yechieli, Yoseph, 1999, Geochemical investigations, in Bear, Jacob, and others, eds., Seawater intrusion in coastal aquifers—Concepts, methods and practices: Dordrecht, The Netherlands, Kluwer Academic Publishers, p. 51–71.

Kindinger, J.L., Davis, J.B., and Flock, J.G., 2000, Subsurface characterization of selected water bodies in the St. Johns River Water Management District, northeastern Florida: U.S. Geological Survey Open-File Report 00–180, 46 p.

Klein, Howard, Armbruster, J.T., McPherson, B.F., and Freiberger, H.J., 1975, Water and the south Florida environment: U.S. Geological Survey Water-Resources Investigation Report 24–75, 165 p.

Knobel, L.L., Chapelle, F.H., and Meisler, Harold, 1998, Geochemistry of the northern Atlantic Coastal Plain aquifer system: U.S. Geological Survey Professional Paper 1404–L, 57 p.

Kohout, F.A., 1964, The flow of fresh water and salt water in the Biscayne aquifer of the Miami area, Florida: U.S. Geological Survey Water-Supply Paper 1613–C, p. 12–32.

Kohout, F.A., 1965, A hypothesis concerning cyclic flow of salt water related to geothermal heating in the Floridan aquifer: New York Academy of Sciences, series II, v. 28, no. 2, p. 249–271.

Kohout, F.A., ed., 1970, Saline water—A valuable resource: Water Resources Research, v. 6, no. 5, p. 1441–1531.

Konikow, L.F., and Reilly, T.E., 1999, Groundwater modeling, in Delleur, J.W., ed., The handbook of groundwater engineering: Boca Raton, Fla., CRC Press, p. 20–1—20–40.

Koszalka, E.J., 1995, Delineation of saltwater intrusion in the Biscayne aquifer, eastern Broward County, Florida, 1990: U.S. Geological Survey Water-Resources Investigations Report 93–4164, 1 sheet.

Krause, R.E, and Clarke, J.S., 2001, Coastal ground water at risk—Saltwater contamination at Brunswick, Georgia, and Hilton Head Island, South Carolina: U.S. Geological Survey Water-Resources Investigations Report 01–4107, 1 sheet.

Krause, R.E., and Randolph, R.B., 1989, Hydrology of the Floridan aquifer system in southeast Georgia and adjacent parts of Florida and South Carolina: U.S. Geological Survey Professional Paper 1403–D, 65 p.

Krieger, R.A., Hatchett, J.L., and Poole, J.L., 1957, Preliminary survey of the saline-water resources of the United States: U. S. Geological Survey Water-Supply Paper 1374, 172 p.

Lacombe, P.J., and Carleton, G.B., 1992, Saltwater intrusion into fresh ground-water supplies, Southern Cape May County, New Jersey, 1890–1991, in Borden, R.C., and Lyke, W.L., eds., Future availability of groundwater resources: American Water Resources Association Technical Publication Series TPS–92–1, p. 287–297.

Lacombe, P.J., and Carleton, G.B., 2002, Hydrogeologic framework, availability of water supplies, and saltwater intrusion, Cape May County, New Jersey: U.S. Geological Survey Water-Resources Investigations Report 01–4246, 151 p.

Lacombe, P.J., and Rosman, Robert, 1997, Water levels in, extent of freshwater in, and water withdrawal from eight major confined aquifers, New Jersey Coastal Plain, 1993: U.S. Geological Survey Water-Resources Investigations Report 96–4206, 8 pls.

Lacombe, P.J., and Rosman, Robert, 2001, Water levels in, extent of freshwater in, and water withdrawals from ten confined aquifers, New Jersey and Delaware Coastal Plain, 1998: U.S. Geological Survey Water-Resources Investigations Report 00–4143, 8 pls.

Landmeyer, J.E., and Belval, D.L., 1996, Water-chemistry and chloride fluctuations in the Upper Floridan aquifer in the Port Royal Sound area, South Carolina, 1917–93: U.S. Geological Survey Water-Resources Investigations Report 96–4102, 106 p.

Langevin, C.D., 2001, Simulation of ground-water discharge to Biscayne Bay, southeastern Florida: U.S. Geological Survey Water-Resources Investigations Report 00–4251, 127 p.

Leahy, P.P., and Martin, Mary, 1993, Geohydrology and simulation of ground-water flow in the northern Atlantic Coastal Plain aquifer system: U.S. Geological Survey Professional Paper 1404–K, 81 p.

LeBlanc, D.R., Guswa, J.H., Frimpter, M.H., and Londquist, C.J., 1986, Ground-water resources of Cape Cod, Massachusetts: U.S. Geological Survey Hydrologic Investigations Atlas 692, 4 sheets.

Leve, G.W., 1983, Relation of concealed faults to water quality and the formation of solution features in the Floridan aquifer, northeastern Florida, USA: Journal of Hydrology, v. 61, no. 1–3, p. 251–264.

Lohman, S.W., and others, 1972, Definitions of selected ground-water terms—Revisions and conceptual refinements: U.S. Geological Survey Water-Supply Paper 1988, 21 p.

Lusczynski, N.J., and Swarzenski, W.V., 1966, Salt-water encroachment in southern Nassau and southeastern Queens Counties, Long Island, New York: U.S. Geological Survey Water-Supply Paper 1613–F, 76 p.

Manheim, F.T., and Horn, M.K., 1968, Composition of deeper subsurface waters along the Atlantic continental margin: Southeastern Geology, v. 9, no. 4, p. 215–236.

Manheim, F.T., Krantz, D.E., Snyder, D.S., Bratton, J.F., White, E.A., Madsen, J.A., and Sturgis, Brian, 2001, Streaming resistivity surveys and core drilling define groundwater discharge into coastal bays of the Delmarva Peninsula: Geological Society of America, Abstracts with Programs, v. 33, no. 6, p. A–42.

Manheim, F.T., Krantz, D.E., Snyder, D.S., and Sturgis, Brian, 2002, Streamer resistivity surveys in Delmarva coastal bays, in Symposium on the Application of Geophysics to Environmental and Engineering Problems (SAGEEP 2002), Las Vegas, Nev., February 10–14, 2002, Proceedings: Las Vegas, Nev., Environmental and Engineering Geophysical Society, Paper 13GSL5, 17 p.

Marella, R.L., 1999, Water withdrawals, use, discharge, and trends in Florida, 1995: U.S. Geological Survey Water-Resources Investigations Report 99–4002, 90 p.

Maslia, M.L., and Prowell, D.C., 1990, Effect of faults on fluid flow and chloride contamination in a carbonate aquifer system: Journal of Hydrology, v. 115, no. 1–4, p. 1–49.

McCobb, T.D., and Weiskel, P.K., 2003, Long-term hydrologic monitoring protocol for coastal ecosystems: U.S. Geological Survey Open-File Report 02–497, 94 p.

McPherson, B.F., and Halley, Robert, 1996, The South Florida environment—A region under stress: U.S. Geological Survey Circular 1134, 61 p.

Meisler, Harold, 1989, The occurrence and geochemistry of salty ground water in the northern Atlantic Coastal Plain: U.S. Geological Survey Professional Paper 1404–D, 51 p.

Meisler, Harold, Leahy, P.P., and Knobel, L.L., 1985, Effect of eustatic sea-level changes on saltwater-freshwater relations in the northern Atlantic Coastal Plain: U.S. Geological Survey Water-Supply Paper 2255, 28 p.

Merritt, M.L., 1985, Subsurface storage of freshwater in south Florida—A digital model analysis of recoverability: U.S. Geological Survey Water-Supply Paper 2261, 44 p.

Merritt, M.L., 1996, Assessment of saltwater intrusion in southern coastal Broward County, Florida: U.S. Geological Survey Water-Resources Investigations Report 96–4221, 131 p.

Metcalf & Eddy, Inc., 1996, Water supply plan submitted to the City of Cape May, New Jersey, April 16, 1996: Somerville, N.J., variously paginated.

Metz, P.A., and Brendle, D.L., 1996, Potential for water-quality degradation of interconnected aquifers in west-central Florida: U.S. Geological Survey Water-Resources Investigations Report 96–4030, 54 p.

Meyer, F.W., 1989, Hydrogeology, ground-water movement, and subsurface storage in the Floridan aquifer system in southern Florida: U.S. Geological Survey Professional Paper 1403–G, 59 p.

Miller, J.A., 1986, Hydrogeologic framework of the Floridan aquifer system in Florida and in parts of Georgia, Alabama, and South Carolina: U.S. Geological Survey Professional Paper 1403–B, 91 p.

Miller, J.A., 1990, Ground water atlas of the United States, segment 6, Alabama, Florida, Georgia, and South Carolina: U.S. Geological Survey Hydrologic Investigations Atlas 730–G, 28 p.

Miller, J.A., 1999, Ground water atlas of the United States, introduction and national summary: U.S. Geological Survey Hydrologic Atlas 730–A, 15 p.

Mitsch, W.J., and Gosselink, J.G., 1993, Wetlands, 2d ed.: New York, Van Nostrand Reinhold, 722 p.

Moore, W.S., 1996, Large groundwater inputs to coastal waters revealed by 226Ra enrichments: Nature, v. 380, no. 6575, p. 612–614.

Moore, W.S., 1999, The subterranean estuary—A reaction zone of ground water and sea water: Marine Chemistry, v. 65, p. 111–125.

National Atmospheric Deposition Program, 2001, National Atmospheric Deposition Program 2000 annual summary: National Atmospheric Deposition Program Data Report 2001–01, 16 p.

National Oceanic and Atmospheric Administration, 1998, Population—Distribution, density and growth, in State of the Coast Report: National Oceanic and Atmospheric Administration, accessed February 15, 2000, at URL *http://state_of_coast.noaa.gov/bulletins/html/pop_01/ appendices/appafull.html*

National Oceanic and Atmospheric Administration, 2001, Relative sea level trends, accessed on April 7, 2001, at URL *http://www.co- ops.nos.noaa.gov/seatrnds.html*

National Research Council, 2000, Clean coastal waters—Understanding and reducing the effects of nutrient pollution: Washington, D.C., National Academy Press, 405 p.

National Research Council, 2001, Aquifer storage and recovery in the Comprehensive Everglades Restoration Plan: Washington, D.C., National Academy Press, 58 p.

Navoy, A.S., 1991, Aquifer-estuary interaction and vulnerability of ground-water supplies to sea-level-rise driven saltwater intrusion: State College, Pennsylvania State University, Ph.D. dissertation, 225 p.

Navoy, A.S., and Carleton, G.B., 1995, Ground-water flow and future conditions in the Potomac-Raritan-Magothy aquifer system, Camden area, New Jersey: New Jersey Geological Survey Report GSR 38, 184 p.

Nemickas, Bronius, and Koszalka, E.J., 1982, Geohydrologic appraisal of water resources of the South Fork, Long Island, New York: U.S. Geological Survey Water-Supply Paper 2073, 55 p.

Nowicki, B.L., Requintina, Edwin, Van Keuren, Donna, and Portnoy, John, 1999, The role of sediment denitrification in reducing groundwater-derived nitrate inputs to Nauset Marsh estuary, Cape Cod, Massachusetts: Estuaries, v. 22, no. 2A, p. 245–259.

Nuttle, W.K., and Portnoy, J.W., 1992, Effect of rising sea level on runoff and groundwater discharge to coastal ecosystems: Estuarine, Coastal and Shelf Science, v. 34, no. 2, p. 203–212.

Odum, J.K., Stephenson, W.J., Williams, R.A., Pratt, T.L., Toth, D.J., and Spechler, R.M., 1999, Shallow high-resolution seismic-reflection imaging of karst structures within the Floridan aquifer system, northeastern Florida: Journal of Environmental and Engineering Geophysics, v. 4, issue 4, p. 251–161.

Olcott, P.G., 1995, Ground water atlas of the United States, segment 12, Connecticut, Maine, Massachusetts, New Hampshire, New York, Rhode Island, Vermont: U.S. Geological Survey Hydrologic Atlas 730–M, 28 p.

Oude Essink, G.H.P., 1999, Impact of sea level rise in the Netherlands, in Bear, Jacob, and others, eds., Seawater intrusion in coastal aquifers—Concepts, methods and practices: Dordrecht, The Netherlands, Kluwer Academic Publishers, p. 507–530.

Parker, G.G., Ferguson, G.E., Love, S.K., and others, 1955, Water resources of southeastern Florida with special reference to the geology and ground water of the Miami area: U.S. Geological Survey Water-Supply Paper 1255, 965 p.

Phelps, G.G., 2001, Geochemistry and origins of mineralized waters in the Floridan aquifer system, northeastern Florida: U.S. Geological Survey Water-Resources Investigations Report 01–4112, 64 p.

Phelps, G.G., and Spechler, R.M., 1997, The relation between hydrogeology and water quality of the lower Floridan aquifer in Duval County, Florida, and implications for monitoring movement of saline water: U.S. Geological Survey Water-Resources Investigations Report 96–4242, 58 p.

Phillips, S.W., 1987, Hydrogeology, degradation of ground-water quality, and simulation of infiltration from the Delaware River into the Potomac aquifers, northern Delaware: U.S. Geological Survey Water-Resources Investigations Report 87–4185, 86 p.

Phillips, S.W., and Ryan, B.J., 1989, Summary of brackish-water intrusion in Coastal Plain aquifers, northern Chesapeake Bay area, Maryland, in Proceedings of ground water issues and solutions in the Potomac River Basin/Chesapeake Bay region: Co-sponsored by The Association of Ground Water Scientists and Engineers, and others, 1989, Washington, D.C., p. 211–235.

Poag, C.W., 1998, The Chesapeake Bay bolide impact—A new view of coastal plain evolution: U. S. Geological Survey Fact Sheet 049–98, 2 p.

Poag, C.W., 1999, Chesapeake invader—Discovering America's giant meteorite crater: Princeton, N.J., Princeton University Press, 183 p.

Poag, C.W., 2000, The Chesapeake Bay impact crater—Meteorite mayhem in Old Virginia: Virginia Explorer, Fall 2000, 9 p.

Poag, C.W., Powars, D.S., Poppe, L.J., and Mixon, R.B., 1994, Meteoroid mayhem in ole Virginny—Source of the North American tektite strewn field: Geology, v. 22, no. 8, p. 691–694.

Pope, D.A., and Gordon, A.D., 1999, Simulation of ground-water flow and movement of the freshwater-saltwater interface in the New Jersey Coastal Plain: U.S. Geological Survey Water-Resources Investigations Report 98–4216, 159 p.

Portnoy, J.W., Nowicki, B.L., Roman, C.T., and Urish, D.W., 1998, The discharge of nitrate-contaminated groundwater from developed shoreline to marsh-fringed estuary: Water Resources Research, v. 34, no. 11, p. 3095–3104.

Powars, D.S., 2000, The effects of the Chesapeake Bay impact crater on the geologic framework and the correlation of hydrogeologic units of southeastern Virginia, south of the James River: U.S. Geological Survey Professional Paper 1622, 53 p.

Powars, D.S., and Bruce, T.S., 1999, The effects of the Chesapeake Bay impact crater on the geologic framework and the correlation of hydrogeologic units of the Lower York-James Peninsula, Virginia: U.S. Geological Survey Professional Paper 1612, 82 p.

Prinos, S., and Overton, K.B., 2000, Water resources data, Florida, water year 1999—v. 2B, South Florida ground water: U.S. Geological Survey Water-Data Report FL–99–2, 527 p.

Pyne, R.D.G., 1995, Groundwater recharge and wells—A guide to aquifer storage recovery: Boca Raton, Fla., Lewis Publishers, 376 p.

Reese, R.S., 1994, Hydrogeology and the distribution and origin of salinity in the Floridan aquifer system, southeastern Florida: U.S. Geological Survey Water-Resources Investigations Report 94–4010, 56 p.

Reese, R.S., 2000, Hydrogeology and the distribution of salinity in the Floridan aquifer system, southwestern Florida: U.S. Geological Survey Water-Resources Investigations Report 98–4253, 86 p.

Reese, R.S., 2002, Inventory and review of aquifer storage and recovery in southern Florida: U.S. Geological Survey Water-Resources Investigations Report 02–4036, 56 p.

Reese, R.S., and Memberg, S.J., 2000, Hydrogeology and the distribution of salinity in the Floridan aquifer system, Palm Beach County, Florida: U.S. Geological Survey Water-Resources Investigations Report 99–4061, 52 p.

Reilly, T.E., 1993, Analysis of ground-water systems in freshwater-saltwater environments, in Alley, W.M., ed., Regional ground-water quality: New York, Van Nostrand Reinhold, p. 443–469.

Reilly, T.E., and Goodman, A.S., 1985, Quantitative analysis of saltwater-freshwater relationships in ground-water systems—A historical perspective: Journal of Hydrology, v. 80, p. 125–160.

Reilly, T.E., and Goodman, A.S., 1987, Analysis of saltwater upconing beneath a pumping well: Journal of Hydrology, v. 89, no. 3–4, p. 169–204.

Richard, J.K., 1976, Characterization of a bedrock aquifer, Harpswell, Maine: Columbus, The Ohio State University, unpublished M.S. thesis, 144 p.

Richter, B.C., Kreitler, C.W., and Bledsoe, B.E., 1993, Geochemical techniques for identifying sources of ground-water salinization: Boca Raton, Fla., C.K. Smoley (CRC Press, Inc.), 258 p.

Rosenau, J.C., Faulkner, G.L., Hendry, C.W., Jr., and Hull, R.W., 1977, Springs of Florida: Florida Geological Survey Bulletin 31 (revised), 461 p.

Rosman, Robert, Lacombe, P.J., and Storck, D.A., 1995, Water levels in major artesian aquifers of the New Jersey Coastal Plain, 1988: U.S. Geological Survey Water-Resources Investigations Report 95–4060, 74 p.

Sanford, W.E., Whitaker, F.F., Smart, P.L., and Jones, Gareth, 1998, Numerical analysis of seawater circulation in carbonate platforms—I. Geothermal convection: American Journal of Science, v. 298, p. 801–828.

Schmerge, D.L., 2001, Distribution and origin of salinity in the surficial and intermediate aquifer systems, southwestern Florida: U.S. Geological Survey Water-Resources Investigations Report 01–4159, 41 p.

Sherif, M.M., 1999, Nile Delta aquifer in Egypt, in Bear, Jacob, and others, eds., Seawater intrusion in coastal aquifers—Concepts, methods and practices: Dordrecht, The Netherlands, Kluwer Academic Publishers, p. 559–590.

Sherif, M.M., and Singh, V.P., 1999, Effect of climate change on sea water intrusion in coastal aquifers: Hydrological Processes, v. 13, p. 1277–1287.

Simmons, G.M., Jr., 1992, Importance of submarine ground-water discharge (SGWD) and seawater cycling to material flux across sediment/water interfaces in marine environments: Marine Ecology Progress Series, v. 84, p. 173–184.

Snow, M.S., 1990, Geochemical determination of salinity sources in surficial and bedrock aquifers of Maine: Orono, University of Maine at Orono, unpublished M.S. thesis, 114 p.

Sonenshein, R.S., 1997, Delineation and extent of saltwater intrusion in the Biscayne aquifer, eastern Dade County, Florida, 1995: U.S. Geological Survey Water-Resources Investigations Report 96–4285, 1 sheet.

Sonenshein, R.S., and Koszalka, E.J., 1996, Trends in water-table altitude (1984–93) and saltwater intrusion (1974–93) in the Biscayne aquifer, Dade County, Florida: U.S. Geological Survey Open-File Report 95–705, 2 sheets.

Sorek, Shaul, and Pinder, G.F., 1999, Survey of computer codes and case histories, in Bear, Jacob, and others, eds., Seawater intrusion in coastal aquifers—Concepts, methods and practices: Dordrecht, The Netherlands, Kluwer Academic Publishers, p. 399–461.

Southwest Florida Water Management District, 2002, Artesian well plugging annual work plan 2002: Brooksville, Southwest Florida Water Management District Quality of Water Improvement Program, 86 p.

Spechler, R.M., 1994, Saltwater intrusion and quality of water in the Floridan aquifer system, northeastern Florida: U.S. Geological Survey Water-Resources Investigations Report 92–4174, 76 p.

Spechler, R.M., 2001, The relation between structure and saltwater intrusion in the Floridan aquifer system, northeastern Florida, in Kuniansky, E.L., ed., U.S. Geological Survey Karst Interest Group, Proceedings, St. Petersburg, Fla., February 13–16, 2001: U.S. Geological Survey Water-Resources Investigations Report 01–4011, p. 25–29.

Spitz, F.J., 1998, Analysis of ground-water flow and saltwater encroachment in the shallow aquifer system of Cape May County, New Jersey: U.S. Geological Survey Water-Supply Paper 2490, 51 p.

Sprinkle, C.L., 1989, Geochemistry of the Floridan aquifer system in Florida and in parts of Georgia, South Carolina, and Alabama: U.S. Geological Survey Professional Paper 1403–I, 105 p.

Stewart, M.T., 1999, Geophysical investigations, in Bear, Jacob, and others, eds., Seawater intrusion in coastal aquifers—Concepts, methods and practices: Dordrecht, The Netherlands, Kluwer Academic Publishers, p. 9–50.

Sun, R.J., and Johnston, R.H., 1994, Regional aquifer-system analysis program of the U.S. Geological Survey, 1978–1992: U.S. Geological Survey Circular 1099, 126 p.

Swarzenski, P.W., and Reich, C.D., 2000, Re-examining the submarine spring at Crescent Beach, Florida: U.S. Geological Survey Open-File Report 00–158, 4 p.

Swarzenski, P.W., Reich, C.D., Spechler, R.M., Kindinger, J.L., and Moore, W.S., 2001, Using multiple geochemical tracers to characterize the hydrogeology of the submarine spring off Crescent Beach, Florida: Chemical Geology, v. 179, p. 187–202.

Task Committee on Saltwater Intrusion, 1969, Saltwater intrusion in the United States: Journal of the Hydraulics Division, American Society of Civil Engineers, v. 95, no. HY5, p. 1651–1669.

Tepper, D.H., 1980, Hydrogeologic setting and geochemistry of residual periglacial Pleistocene seawater in wells in Maine: Orono, University of Maine at Orono, unpublished M.S. thesis, 126 p.

Trapp, Henry, Jr., and Meisler, Harold, 1992, The regional aquifer system underlying the northern Atlantic Coastal Plain in parts of North Carolina, Virginia, Maryland, Delaware, New Jersey, and New York—Summary: U.S. Geological Survey Professional Paper 1404–A, 33 p.

Urish, D.W., and Qanbar, E.K., 1997, Hydrologic evaluation of groundwater discharge, Nauset Marsh, Cape Cod National Seashore, Massachusetts: U.S. National Park Service, New England System Support Office, Technical Report NPS/NESO-RNR/NRTR/97–07, 69 p.

U.S. Army Corps of Engineers and South Florida Water Management District, 1999, Central and Southern Florida Project Comprehensive Review Study: Final Integrated Feasibility Report and Programmatic Environment Impact Statement, 27 p.

U.S. Bureau of the Census, 1996, Census of population and housing, 1995: Public Law (P.L.) 94–171 Data Technical Documentation, prepared by the Bureau of the Census, Washington, D.C., data accessed from U.S. Geological Survey, February 15, 2000, at URL *http://water.usgs.gov/watuse/spread95.html*

U.S. Environmental Protection Agency, 1992, Secondary drinking water regulations—Guidance for nuisance chemicals: U.S. Environmental Protection Agency Report EPA 810/K–92–001, accessed September 22, 2000, at URL *http://www.epa.gov/safewater/consumer/2ndstandards.html*

U.S. Environmental Protection Agency, 2002a, 2002 Edition of the drinking water standards and health advisories: U.S. Environmental Protection Agency Report EPA 822–R–02–038, accessed March 13, 2003, at URL *http://www.epa.gov/waterscience/drinking/standards/dwstandards.pdf*, 19 p.

U.S. Environmental Protection Agency, 2002b, Drinking water advisory—Consumer acceptability advice and health effects analysis on sodium: U.S. Environmental Protection Agency EPA 822–R–02–032 (April 2002), accessed March 13, 2003, at URL *http://www.epa.gov/safewater/ccl/pdf/sodium.pdf*, 34 p.

U.S. Geological Survey, 1996, U.S. Geological Survey Programs in New Jersey: U.S. Geological Survey Fact Sheet 030–96, 4 p.

U.S. Geological Survey, 1999, The quality of our Nation's waters—Nutrients and pesticides: U.S. Geological Survey Circular 1225, 82 p.

U.S. Geological Survey, 2000, National water-use data files, 1995 estimated water use in the United States, county data files: U.S. Geological Survey data available on the World Wide Web, accessed February 15, 2000, at URL *http://water.usgs.gov/watuse/spread95.html*

van Dam, J.C., 1999, Exploitation, restoration, and management, *in* Bear, Jacob, and others, eds., Seawater intrusion in coastal aquifers—Concepts, methods and practices: Dordrecht, The Netherlands, Kluwer Academic Publishers, p. 73–125.

Voronin, L.M., Spitz, F.J., and McAuley, S.D., 1996, Evaluation of saltwater intrusion and travel time in the Atlantic City 800-foot sand, Cape May County, New Jersey, 1992, by use of a coupled-model approach and flow-path analysis: U.S. Geological Survey Water-Resources Investigations Report 95–4280, 27 p.

Weiskel, P.K., DeSimone, L.A., and Howes, B.L., 1996, A nitrogen-rich septage-effluent plume in a coastal aquifer, marsh, and creek system, Orleans, Massachusetts—Project summary, 1988–95: U.S. Geological Survey Open-File Report 96–11, 20 p.

Wrege, B.M., and Daniel, C.C., III, 2001, Use of cores, borehole geophysical logs, and high-resolution seismic-reflection data to delineate paleochannels underlying the U.S. Marine Corps Air Station, Cherry Point, North Carolina [abs.], *in* 2001 Abstracts with Programs, 50th Annual Meeting, Southeastern Section, April 5–6, 2001: Raleigh, N.C., The Geological Society of America, v. 33, no. 2, p. A–25.

Yobbi, D.K., 1997, Simulation of subsurface storage and recovery of effluent using multiple wells, St. Petersburg, Florida: U.S. Geological Survey Water-Resources Investigations Report 97–4024, 30 p.